Anailton Sales de Melo
Flavio Justino

Wildfires Under Current Climatic Conditions and Global Warming

Anailton Sales de Melo
Flavio Justino

Wildfires Under Current Climatic Conditions and Global Warming

Susceptibility of the Brazilian environment to wildfires

ScienciaScripts

Imprint

Cover image: www.ingimage.com

This book is a translation from the original published under ISBN 978-3-330-75968-8.

Publisher:
Sciencia Scripts
is a trademark of
Dodo Books Indian Ocean Ltd. and OmniScriptum S.R.L publishing group

120 High Road, East Finchley, London, N2 9ED, United Kingdom
Str. Armeneasca 28/1, office 1, Chisinau MD-2012, Republic of Moldova, Europe
Printed at: see last page
ISBN: 978-620-8-31059-2

Summary

BIOGRAPHY

Anailton Sales de Melo, son of José Milton Castelo Branco de Mélo and Anaires Clementino de Mélo, was born in Campina Grande, Paraiba on May 19, 1975.

In September 1996, he graduated in Physics from the State University of Paraiba (UEPB), writing a Course Conclusion Paper (TCC) entitled, Physics textbooks and their contributions to the teaching-learning process, in Campina Grande, where he was a scholarship holder and monitor. He became a specialist in Physics Teaching at the University of São Paulo in 2000. He worked as a high school physics teacher in the city of São Paulo for the São Paulo State Network from 1999 to 2007.

He has also worked as a Physics Teacher for the State of Minas Gerais, teaching high school Physics. He began his master's degree in Agricultural Meteorology, concentrating on Climatology, at the Federal University of Viçosa (UFV) in August 2007 and finished in March 2009.

He has been teaching in the Pernambuco State Network since 2014, teaching Physics, Chemistry and Mathematics at the Full-Time Reference School.

Chapter 1

1.0 INTRODUCTION

People have long been aware of the close relationship between man and the climate due to the influence of climatic variations on social well-being. With the increase in consumption needs associated with the rise in population over the last 10,000 years, agricultural activities have grown. Agricultural activities, in turn, require seasonal soil management, which in most cases involves the use of fire. According to Laturner and Scherer (2004), man's evolutionary history has seen him use fire from the earliest times. In recent decades, various sectors of society have become increasingly concerned about the indiscriminate use of fire. Cultural fire, as Agee (1993) calls ignitions of human origin, is, however, just as important, if not more so, than natural fire.

The burning of biomass in ecosystems due to the expansion of the agricultural frontier, the conversion of forests and savannahs into pastures, and the renewal of agricultural crops, are some of the most important factors causing impacts on climate and biodiversity (KIRCHOFF, 1997; COSTA *et al.,* 2007). Fires also cause soil depletion, the destruction of vegetation, erosion problems, atmospheric pollution and are linked to changes in the chemical composition of the atmosphere (CRUTZEN and ANDRAEA, 1990).

Wildfires and forest fires have not always been directly attributed to human action. The occurrence of fire depends on a number of factors that allow the combustion reaction to begin and its continuation depends mainly on the potential energy stored in the combustible material (RIBEIRO and SOARES, 1998). For example, palynological studies in the central region of the Brazilian cerrado indicate the presence of charcoal in reconstructions dating back 32,000 years before the present (AP) (FERRAZ-VICENTINI, 1999; SALGADO-LABOURIAU and FERRAZ-VICENTINI, 1994). This period, which is part of the last ice age, which lasted from 100,000 to 10,000 years BC, is characterized by a colder and drier climate than today (JUSTINO, 2004; CRUZ *et al.,* 2005). However, very little is known about the causes that led to fire-prone conditions during this period. It is believed that the most likely cause of these glacial fires was the incidence of lightning.

In the Amazon region, the presence of charcoal fragments in rainforest soils has been studied by several authors over the last 15 years (SANFORD *et al.*, 1985; SALDARRIAGA and WEST, 1986), with dates of fragments ranging from 250 to 6500 years BP.

Climate changes, such as those caused by *El Nino* (PHILANDER, 2001), can contribute to increasing the potential risk of burn-offs and accidental fires in Brazil. The effects of *El Nino* anticipate agricultural burn-offs, causing an extension of the traditional period and, consequently, an increase in the occurrence of burn-offs and forest fires throughout the year.

The role of fire associated with deforestation is widely known, but its impacts are still poorly documented. Every year accidental under-fires, which have been called hidden deforestation, consume a large part of the Amazon rainforest (NEPSTAD *et al.*, 1999). Fire in vegetation, after agricultural and urban activities, is the factor that contributes most to global deforestation/forestation trends (BOND and VAN WILGEN, 1996).

Post-burning environmental conditions reinforce behaviors that favor the emergence of new fires (RIBEIRO and BONFIM, 2000). The trees removed during deforestation are often large. Their removal leaves forests vulnerable, resulting in greater radiation and, consequently, higher temperatures, in other words, altering their entire microclimate and favoring greater susceptibility to future fires (Fig. 1.1). Among the risks associated with cutting down large areas of forest is the modification of agents in the hydrological cycle, such as the reduction in evapotranspiration (COOK and VIZY, 2008) and, consequently, the re-feeding of atmospheric water vapor, which can reduce precipitation (COSTA and FOLEY, 2000).

Figure 1: Deforestation

It should also be noted that, as shown by Andreae *et al.* (2004), the greater incidence of aerosols in the atmosphere due to the presence of fires is an unfavorable factor for precipitation because of their influence on reducing the size of cloud droplets. Small droplets are carried by rising air currents to higher levels in the atmosphere, which leads to a lack of rainfall and, consequently, encourages new fires.

These factors certainly lead to negative impacts on biodiversity, as they decrease the amount of biomass stored in forests, reducing the amount of water released from the soil that returns to the atmosphere through the vegetation (KLAR, 1984). The interaction between fires and climate is also associated with changes in the local energy balance, since the particles emitted by fires interact with solar radiation, affecting the atmospheric radiative balance (RAMANATHAN *et al.*, 2001; KAUFMAN *et al.*, 2002; PROCÓPIO *et al.*, 2004). In addition, due to the scattering properties of the particles emitted by fires, there is a reduction in the ratio between direct and diffuse radiation reaching the tree canopy (YAMASOE *et al.*, 2005) (Fig. 1.2).

Figure 1.1: Smoke

The impacts of burning are highlighted in issues involving future climate change, especially the intensification of Global Warming (GW). Biomass burning plays an important role in global emissions of carbon and other trace gases, as well as particulate matter (ANDREAE *et al.,* 1996a,b; CRUTZEN and ANDREAE, 1990). Burning at a global level is believed to be the second largest source of greenhouse gas emissions. For example, the amount of CO_2 released from fires in Indonesia in 1997 and 1998 was equivalent to 25% of the total CO_2 emitted annually from burning fossil fuels (PAGE *et al.,* 2002). In Brazil, fires are responsible for around 60-70% of carbon dioxide (CO_2) emissions into the atmosphere, i.e. it is the main source of emissions.

In addition to CO_2 emissions, fires are also an important emitter of carbon monoxide (CO) into the atmosphere (OLIVEIRA, 2005). The amount of this gas emitted in a fire depends on the characteristics of the fire: fast fires with a good supply of oxygen and which reach high temperatures produce little CO, while inefficient fires produce large amounts of this gas as a product of incomplete combustion. According to Andreae (1991), most fires occur in developing countries in the tropical region. These countries are responsible for 87% of global emissions from burning, estimated at 3940 Tg[C]/year.

Currently, the highest incidence of fires in Brazil occurs between August and

November, with a peak in September (SISMANOGLU and SETZER, 2002; JUSTINO *et al.,* 2002). This seasonal peak is directly associated with the dry period that precedes the rainy season and planting (Fig. 1.2).

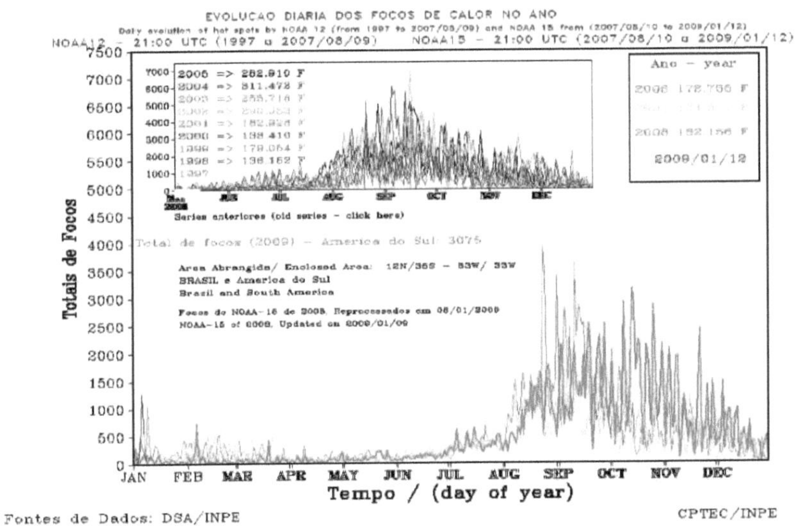

Figure 1.2: Temporal evolution of total fires (2005-2008) Source: Center for Weather Forecasting and Climate Studies (CPTEC)

In the far north of Brazil, on the other hand, fires are concentrated between January and March. As shown in Figure 1.3, most of Brazil is affected by fires. It can also be seen that the states of Mato Grosso (MT), Para (PA), Sâo Paulo (SP), Roraima (RO) and Maranhâo (MA) concentrate most of the fires. In 2005, the Western Amazon suffered a prolonged drought which led to large open fires and forest fires, affecting more than 400,000 hectares in eastern Acre (BROWN *et al.*, 2006a, b).

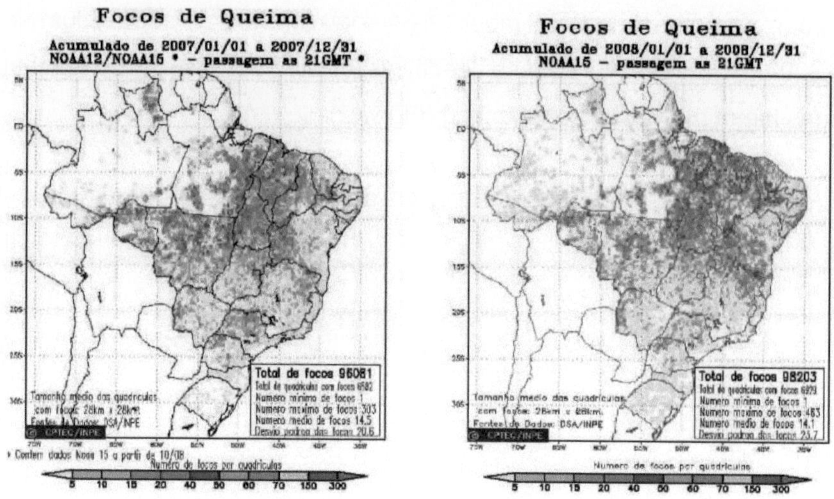

Figure 1.3: Spatial and temporal distribution of fires in 2007 and 2008 Source: Center for Technological Research (CPETC)

The action of fire on the environment also depends on factors other than those necessary for combustion to occur. These include combustible material, climatic conditions and topography. The action of each of these factors is different for each region and for each time of year, causing great differences in fire behavior (BATISTA, 1995; RIBEIRO and BONFIM, 2000).

Studies on the impact of fire in tropical forests are becoming increasingly important as the effects of fires have negative global repercussions, particularly due to their influence on atmospheric composition and biodiversity stocks (Whitmore, 1990; COCHRANE, 2003). Although studies are currently underway, there are no records in the literature of investigations that have focused on fire risk analysis under other climatic conditions, such as global warming. The application of methodologies that assess fire risk is useful for a spatial extrapolation of the regions most prone to fire in a global warming scenario. In this way, it is worth pointing out that fire increases the flammability of flora and starts a positive feedback loop, in which tropical forests are replaced by vegetation that is even more prone to fire.

Protecting the forest and its associated resources from the devastating effects of fires is the primary motivation for studying fire risk (HAINES *et al.,* 1998). An important step towards reducing the impacts of fires is to investigate the susceptibility of a given environment to burning or even the spread of fire (fire risk). This is already done in some countries in the Northern Hemisphere, especially the United States and Canada. A widely used index, which can be understood as fire risk, is the Haines index - HI (HAINES, 1988). Originally called the low atmospheric severity index, the HI is computed from the air temperature and dew point in the lower troposphere and characterizes the potential impact that atmospheric conditions have on the spread of fire. A first analysis of this index for Brazil was carried out by Lemos (2006), where it was found that the first half of September is the most critical period in terms of the incidence of hotspots and the maximum risk in central Brazil.

For South America and the Caribbean, the Burn Project team at the CENTER FOR WEATHER FORECASTING AND CLIMATE STUDIES (CPTEC) of the NATIONAL INSTITUTE FOR SPACE RESEARCH (INPE) has developed a methodology that relates atmospheric conditions (temperature, precipitation, humidity) and vegetation cover into an index that quantifies a region's aptitude for fire development (SISMANOGLU *et al.,* 2002). These two methods - iH and CPTEC-RF - will be described in more detail below.

Based on the points discussed above, the aim of this study is to investigate how favorable the environment is to the occurrence of fires (fire risk). The study will be carried out under two different climatic conditions: current and global warming. To this end, data from coupled numerical simulations conducted with the ECHAM5/MPI-OM model will be used. The variables to be used in the study are: maximum temperature (Tmax), precipitation (prec) and dew point temperature (Td), which are output variables of the ECHAM5/MPI-OM model. These numerical simulations come from the "ENSEMBLE SIMULATIONS OF EXTREME WEATHER EVENTS UNDER NONLINEAR CLIMATE **CHANGE**" (ESSENCE) project (http://www.knmi.nl/~sterl/Essence).

Specifically, the aim is to:

1) To identify the main Brazilian regions most susceptible to fires under a global warming scenario as a strategy for adapting to and mitigating climate change.

2) Compare two fire risk methods for the current and future periods.

Chapter 2

2.0 DATA USED

This chapter describes the observational and numerical data sets used, as well as the methodology employed.

2.1 NCEP/NCAR REVIEWS

Meteorological observations are fundamental to weather and climate forecasting. For this reason, special attention should be paid to collecting and processing information on the state of the atmosphere. In view of this, the WORLD METEOROLOGICAL ORGANIZATION (WMO) has standardized the way meteorological data is obtained. The information obtained according to WMO standards can be called conventional observations, as they follow a rigorous process of observation, processing and dissemination of the data collected.

This cooperative project of a reanalysis dataset relies on a Data Assimilation (DA) technique to produce a significant number of weather and climate variables from radio soundings at different altitudes, data from satellites, planes, ships and under the continent. The advantage of this database lies in the fact that it provides a large number of parameters (surface and altitude), all related to the dynamics of the atmosphere, forming a very coherent and up-to-date set of information, which is a very important feature for climate variability studies.

In this study, the analyses are based on 20 climatological years for the period from January 1980 to December 2000, presented in monthly maps. The reliability of the NCEP/NCAR reanalyses for the Southern Hemisphere (HS) has improved significantly since 1979, benefiting from the assimilation of satellite data, which partially completes the shortage of conventional observations (HINES *et al.*, 2000).

According to Pinto (2007), the NCEP/NCAR historical reanalysis series are widely used in studies describing general atmospheric conditions, both for analysis of isolated areas and for global analysis, although they do present the problem of overestimating rainfall in a large part of the tropical zone. Despite these problems, reanalysis data are one of the few that include atmospheric dynamics in their production.

2.2 Simulated Data: ECHAM5/MPI-OM

Description of the ECHAM5/MPI-OM Model

As mentioned above, the proposed questions will be investigated using numerical simulations for the current period and for the Global Warming (GW) scenario. To this end, the sixth generation of the Model of General Atmospheric Circulation (MCGA), the ECHAM5/MPI-OM model, developed at the Max-Planck Institute for Meteorology (MPIM) in Hamburg, Germany, will be used. The two components of the model, namely ECHAM5 for the atmosphere and MPI-OM for the ocean, are well documented (Jungclaus *et al., 2006).* Their coupled version was used for the climate scenarios in the Intergovernmental Panel on Climate Change report (Roeckner *et al.*, 2003). In this context, it should be noted that the configuration used in this study has a triangular truncation at wave number T63, approximately 1.8° of latitude and longitude, which represents a spatial resolution of approximately 180 x 180 km, in the equatorial band, with 31 vertical levels.

The ocean model of the Max Planck Institute (MPI-OM) (Marsland *et al.,* 2003) uses the primitive equations for a hydrostatic fluid with a free surface. The discretization has 40 vertical levels with a horizontal resolution of 1.5°.

The analyses evaluated in this work are based on the climatology for two different periods: the current period, which covers the years 1980 to 2000, and the future period, which is set between 2080 and 2100. This longer interval was chosen because of the need to investigate the risk of fires in conditions that are extremely different from current conditions.

The proposed economic scenario is based on scenario A1B. These climate scenarios project a description of a future state of emissions of the main greenhouse gases and aerosols. For the intermediate A1B scenario, the atmospheric concentration of CO_2 is 775 ppmv (parts per million by volume). Vera *et al.* (2006), using various IPCC models, based on future projections for the period 20702099 of the A1 B scenario, points to an increase in precipitation over subtropical South America (SSA) during the summer, a reduction in precipitation during the winter over almost the entire continent and a reduction in precipitation during the four seasons throughout the year in the southern region of the Andes. In this way, it can be argued that if the proposed socio-economic conditions continue, the future climate scenario could be favorable to greater environmental flammability due to longer periods of drought.

Chapter 3

3.0 CLIMATIC BEHAVIOR

3.1 CLIMATOLOGICAL ANALYSIS

Brazil is a vast continent and has different temperature and precipitation regimes, due to the dynamic and synoptic characteristics that act over South America (SSA) (GRIMM, 2004). It is therefore assumed that the annual cycles can be treated as harmonic terms over the years within a given period. An analysis of the seasonal cycle of temperature (Temp), precipitation (Prec) and relative humidity (RH), based on observed data and/or data from numerical simulations, is essential for diagnosing which climatic variables are decisive and which regions are most vulnerable to fires.

Climate research based on numerical simulations requires an assessment of how well these numerical results reproduce the observed data. To this end, Figures 3a, b, c, d, e, f show the spatial distribution of air temperature, rainfall and relative humidity from NCEP reanalyses and ECHAM5/MPI-OM model data, from June to October between the years 1980 and 2000. It is during the months of June and October that the highest number of fires is detected in Brazil.

According to the NCEP reanalyses, the regions with the highest temperatures from June to October are: Mato Grosso, Tocantins, Rondônia, Mato Grosso do Sul and Goiás (Fig. 3a). This behavior is also verified using ECHAM5 data (Fig. 3b). The ECHAM5 data, however, differs from the NCEP data, as the model shows temperatures up to 5°C higher than the NCEP data. Comparing Figures 3a and 3b, it can be seen that ECHAM5 shows an area with high temperature values between Parà and Amazonas, which does not exist in the reanalyses. It is important to note that along the coast and southeast of Brazil, where there is a greater density of observed data, the model and the reanalyses show greater similarity. The NCEP reanalyses assimilate observed data in their processing.

The spatial distribution of rainfall is shown in Figures (3c, d). It can be seen that, with the exception of the southern region and the northwest of the northern region, rainfall is in the order of 4 mm/day for both data (reanalysis and model). In the region stretching from the west of the Amazon to the southeast, the ECHAM5 data shows higher rainfall due to a better reproduction of the South Atlantic Convergence Zone (SACZ). Although its activity is weakened at this time of year. It should also be noted that both data are coherent for the Brazilian Northeast (NEB), the southern region and the northwest of Brazil.

A similar analysis for relative humidity shows that, according to the NCEP reanalyses, the regions with the lowest relative humidity values are the center-south of Mato Grosso and the western strip of Mato Grosso do Sul (Fig. 3e). In contrast, the ECHAM5 results show the lowest relative humidity values shifted to the center-north of Mato Grosso, the center-south of the state of Parà and part of Tocantins (Fig. 3f). In the other areas, both sets of data show considerable similarity, particularly in southern Brazil. In general terms, the climate modeled by ECHAM5 was found to be warmer, with higher rainfall and lower relative humidity values, when compared to the NCEP reanalyses. It is important to note that the biggest differences between the modeled data and the reanalyses are in regions with substantially low data coverage.

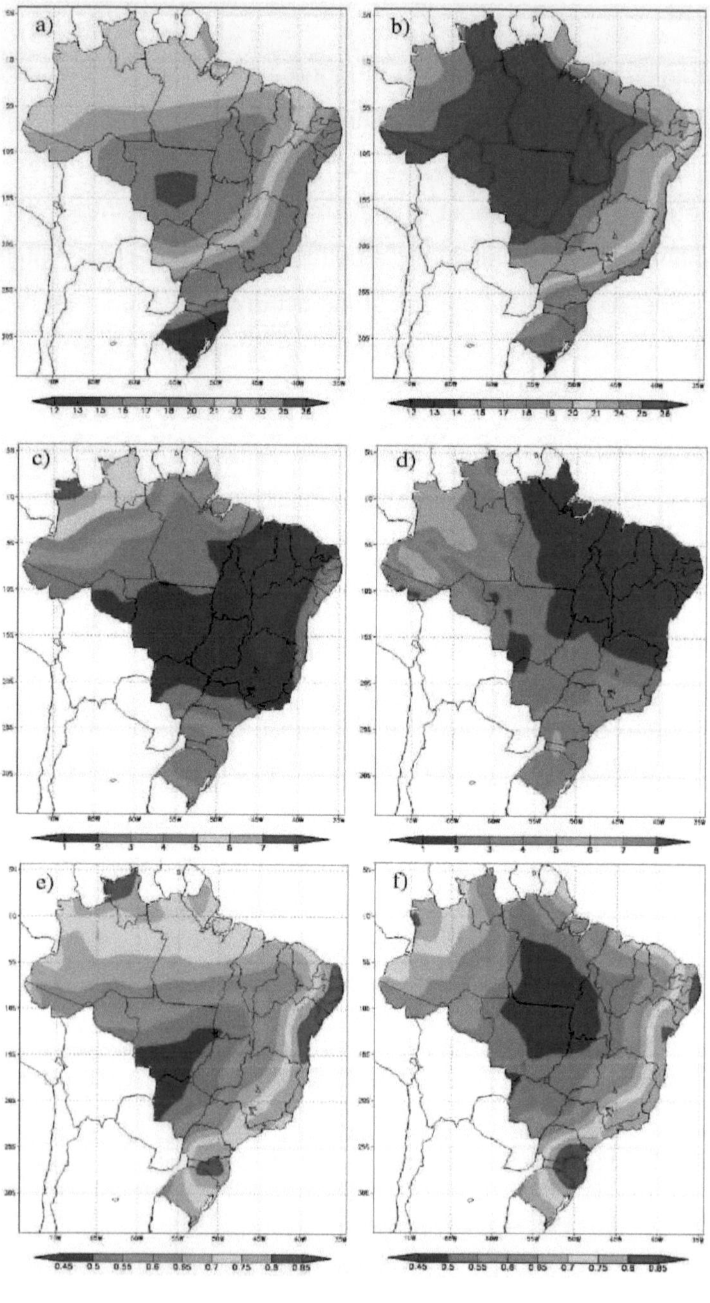

Figure 3: Climatology from June 1980 to October 2000 Reanalyses (NCEP): a) Temperature (°c), c), Precipitation (mm), e) RH (%) Simulations (ECHAM5): b) Temperature (°c), d)

Precipitation (mm), f) RH (%)

3.2 HARMONIC ANALYSIS

Harmonic analysis has been a useful tool in the study of weather patterns, representing the variations of a time series on the basis of periodic functions of sine and cosine (ASLAN and TOPÇU, 1994). These functions present frequencies as multiples of a fundamental frequency. According to Wilks (1995) and Krikyla and Hameed (1989), a given time series of data from equidistant samples of a variable can be represented by a harmonic function:

$$y_t = \bar{y} + \sum_{j=1}^{N/2} C_j \cos(\omega_j t - \phi_j)$$

In which:

$\bar{y}$ = average of the values in the series,

C_j = amplitude;

ω_j = frequency; and

ϕ_j = phase angle.

It is possible to write the previous equation using the sine function. However, the cosine function is generally used, as the phase angle (ϕ) can easily be determined as corresponding to the time where the harmonic function is maximized:

$$t = \frac{\phi_j N}{2\pi}$$

The equations for calculating harmonic amplitude according to (EMERY and THOMSON, 1997) are:

- Amplitude (C):$_j$

$$C_j = \sqrt{A_j^2 + B_j^2}$$

We have: A_j = coefficient of the cosine;

B = jcoefficient of the sine;

C = jvector amplitude.

the terms A and B are obtained by the following formulas:

$$A_j = \frac{2}{N}\sum_{t=1}^{N} y_t \cos\left(\frac{2\pi t}{N}\right)$$

$$B_j = \frac{2}{N}\sum_{t=1}^{N} y_t \operatorname{sen}\left(\frac{2\pi t}{N}\right)$$

e,

where: k = 2πt, t=month (t = 1,2, ... ,12)

- Phase (ϕ)

The phase angle (ψ) is calculated from one of the following equations, depending on the value obtained for (A_j):

$$\phi_j = \begin{cases} \tan^{-1}\frac{B_j}{A_j} & A_j > 0 \\ \tan^{-1}\frac{B_j}{A_j} \pm \pi \text{ou} \pm 180^\circ & A_j < 0 \\ \frac{\pi}{2} \quad \text{ou } 90^\circ & A_j = 0 \end{cases}$$

The contribution of each harmonic (^{j}) **to** the total variance of the data series is calculated from the ratio between the harmonic's amplitude and twice the population variance (PANOFSKY and BRIER, 1968):

$$j = \frac{C_j^2}{2s^2}$$

Harmonic analysis has been used in various studies for a wide variety of purposes. As discussed by Azzali and Menetti (2001), this type of analysis is useful for characterizing different climatic regimes in transitional regions as well as classifying climatic zones. The advantage of using this mathematical approach is the possibility of identifying the dominant climate pattern in the spatial and temporal domain. In this way, it is possible to demonstrate the main characteristics related to seasonality and, in particular, to investigate how these climatic patterns correlate with agricultural activities associated with biomass burning and the risk of fire.

Results based on harmonic analysis express the behavior of a variable as a function of its periodicity, amplitude and phase (period in which the peak in amplitude occurs). In this process, the aim is to examine how the climatic component contributes and interacts to produce the most representative cycle of the observed and modeled data. The harmonic behavior of observed and modeled temperature is shown in Figure 3.1, for precipitation in Figure 3.2, and for relative humidity in Figure 3.3.

3.2.1 Harmonic Temperature Analysis (Temp).

Figures (3.1a, b) show that in a large part of central Brazil, the region where most fires are concentrated, the annual cycle explains around 50% of the variance in the data. This indicates a small annual temperature range. However, the analyses resulting from ECHAM5 show a greater representation of the first harmonic (greater explained variance) over the state of Parà. With regard to the phase field, or months when the maximum temperature values occur, there is a lag of one or two months between the NCEP data and the simulated ECHAM5 data (Fig. 3.1c, d).

The temperature amplitudes shown in Figures (3.1e, f) illustrate that there are small zonal variations, with the greatest variations detected along the latitudes. In both sets of data evaluated, the southern part of Brazil shows the greatest seasonal temperature variations because it has a better definition of the seasons, winter and summer. According to the results proposed by the ECHAM5 model, there is a region with substantial seasonal temperature variation, located over the northern part of Mato Grosso and a large part of Para. This indicates that temperature differences in this area could reach 4°C.

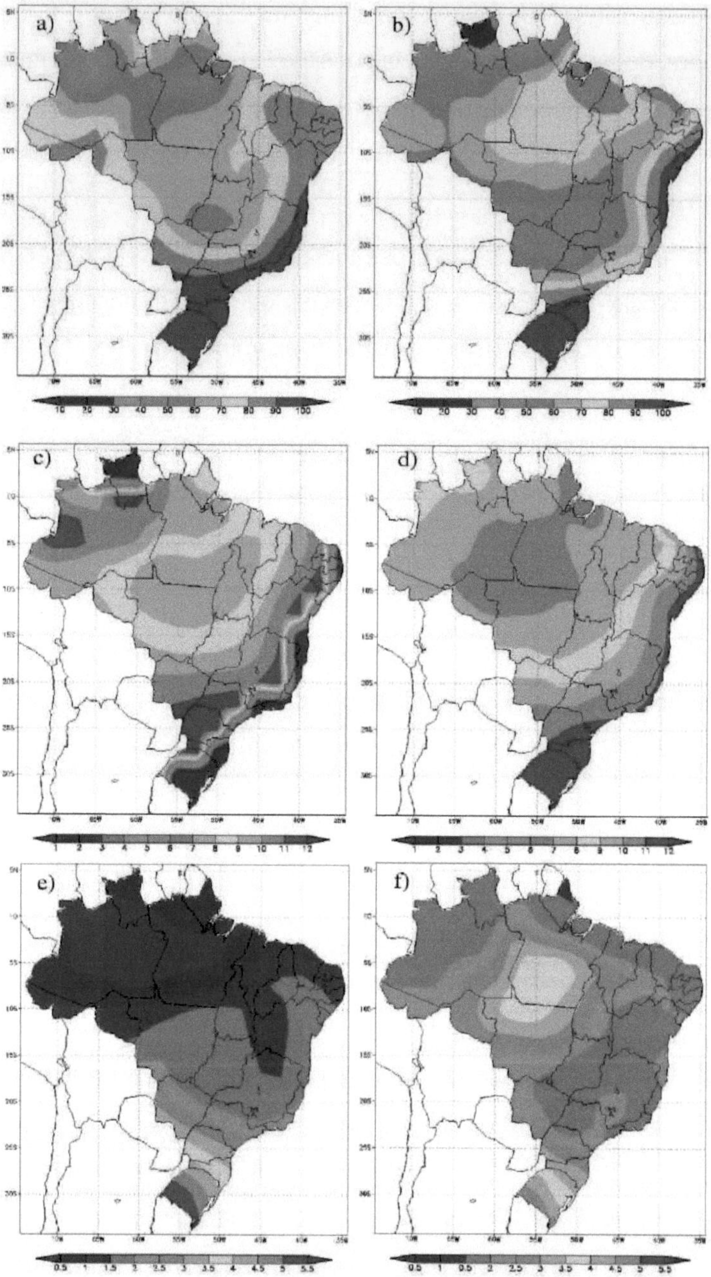

Figura 3.1: First Harmonic Temperature from June 1980 to October 2000 Reanalyses (NCEP): a) Variance (%), c) Phase (month), e) Amplitude (°C) Simulations (ECHAM5): b)

Variance (%), d) Phase (month), f) Amplitude (°C)

3.2.2 Harmonic Analysis of Precipitation (Prec).

Figures (3.2a, c, e) show the variance, phase and amplitude of the first harmonic of precipitation obtained from NCEP reanalysis data, and Figures (3.2b, d, f) show the same fields from simulated ECHAM5 data, respectively. All regions show significant variances between 80%-100% for the NCEP data (Fig. 3.2a, b), except for the southern region (10%-60%). This can be understood as follows: where the variance exceeds 60%, there is in fact a well-defined period of both drought and rainfall. On the other hand, in southern Brazil, where the variance is small, there is no well-defined dry period, i.e. the annual amplitude is small. The same behavior can be seen in the data simulated by ECHAM5. The differences are observed in the northern part of the region with variance values between 30%-80%. This shows that the NCEP data does not capture the seasonal variation in precipitation associated with the meridional movement of the Inter-Tropical Convergence Zone (ITCZ).

Figure (3.2c) shows that in the northern regions and much of the northeast, the rainy season is concentrated between January and April according to NCEP data. For the southern and southeastern regions and the state of Mato Grosso do Sul, the rains occur in the last few months of the year. Figure (3.2d), in turn, shows data from ECHAM5, which has a rainy period with some differences in the months of maximum rainfall, mainly in the south and southeast regions.

The regions with large amplitudes have a great distinction between the rainy and dry seasons. Figure (3.2e) shows the states of Acre, Amapà, Parà, Mato Grosso, Goiàs, Tocantins, Bahia and Maranhào, with amplitudes of between 4mm-6mm. Figure (3.2f), based on ECHAM5 data, shows the highest amplitudes from the south of the northern region to the northeast, and in the states of Mato Grosso, Goiàs and Tocantins, with values between 4mm and 6mm.

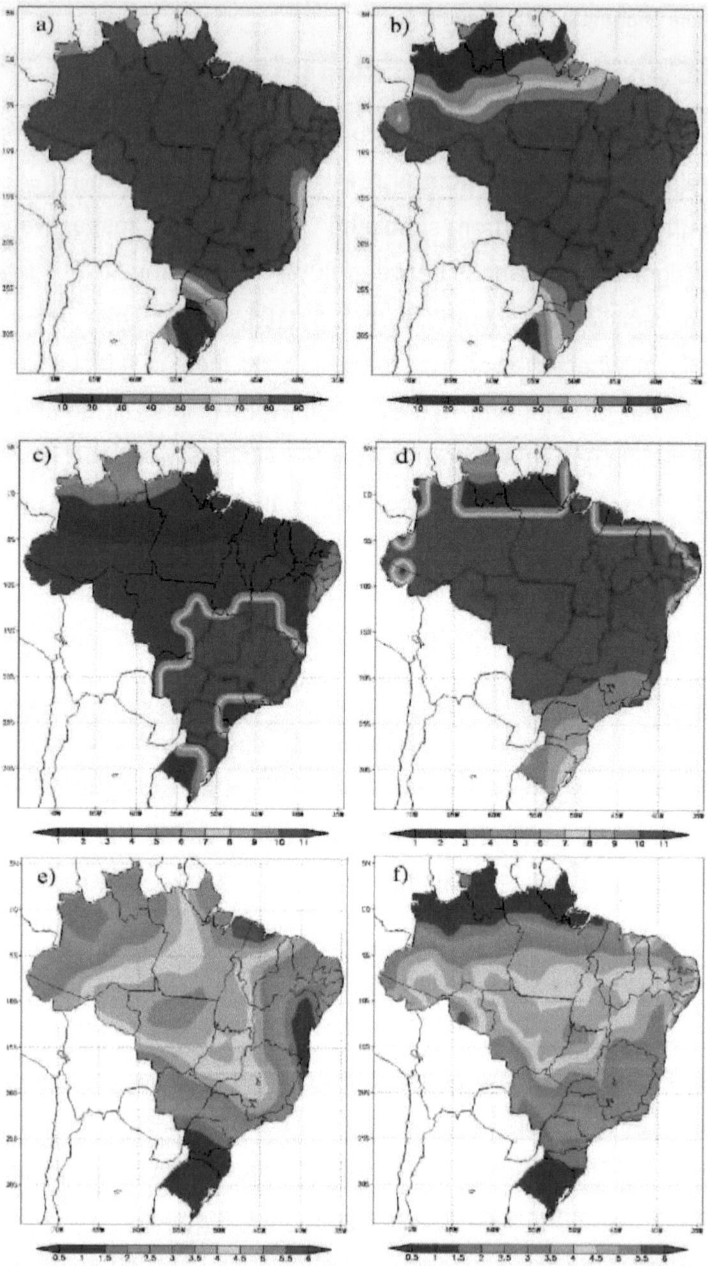

Figura 3.2: First Harmonic Rainfall from June 1980 to October 2000 Reanalyses (NCEP): a) Variance (%), c) Phase (month), e) Amplitude (mm) Simulations (ECHAM5): b) Variance (%), d) Phase (month), f) Amplitude (mm)

3.2.3 Harmonic Analysis of Relative Humidity (RH).

In terms of the variance explained by the first harmonic, both sets of data show values above 80%. Only the states in the southern region and the far north of the northern region show values below 80%. It is important to note the similarity between the data sets.

An analysis of the phase field of the first harmonic of relative humidity for the NCEP data (Fig. 3.3c) shows that relative humidity peaks over central Brazil, Minas Gerais, Rio de Janeiro and São Paulo between October and December. The lows will therefore be observed between May and June. This behavior differs when compared to the ECHAM5 data, where this characteristic is seen only for the states of Mato Grosso do Sul, Goiás, São Paulo, Rio de Janeiro and northern Paranà. The model also simulates that for part of the northern region, Mato Grosso and Minas Gerais, the highest/lowest relative humidity values are reached at the beginning/middle of the year.

In terms of the amplitude of the annual cycle, both sets of data (NCEP and ECHAM5) show a similar spatial distribution with larger amplitudes of between 0.15% and 0.25% in the central-western region, as well as a similarity in the other regions with lower values of around 0.03%-0.12% (Fig. 3.3e, f). It is important to note that the ECHAM5 data shows the state of Parà with high seasonal variation in relative humidity, a characteristic not seen in the NCEP data. High values in the amplitude of relative humidity show a high fluctuation between hot, *dry periods and* cold, wet periods. As will be discussed below, areas of high amplitude can be characterized as fire-prone zones.

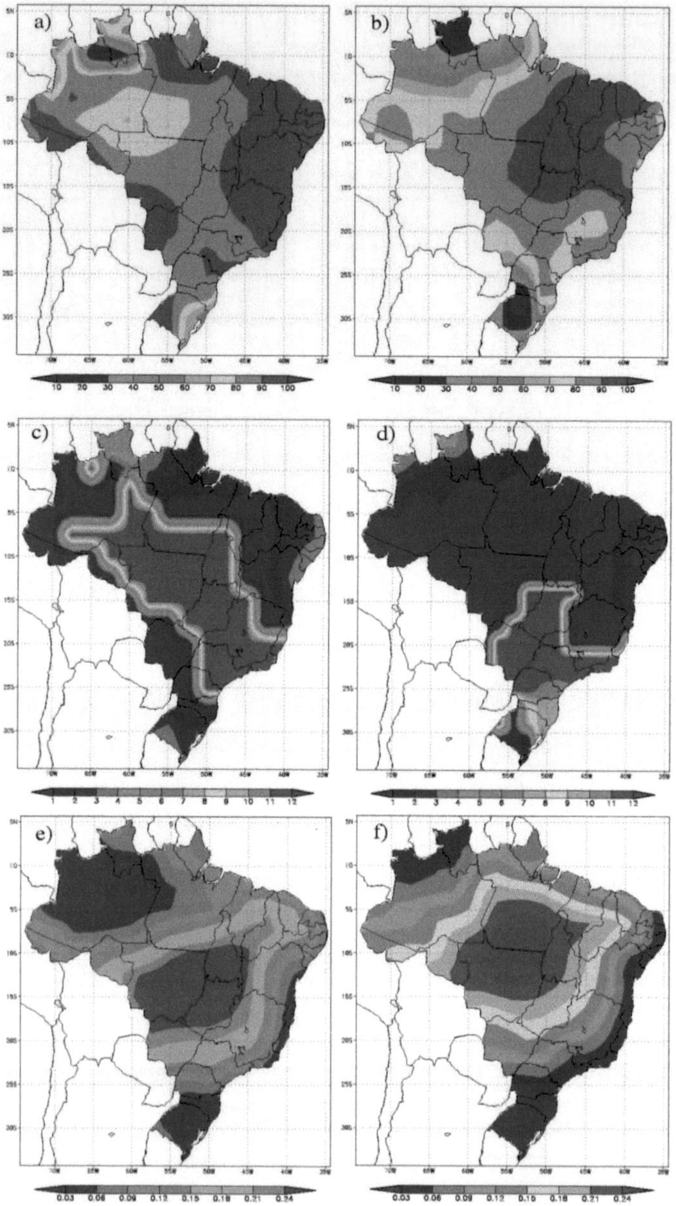

Figura 3.3: First Harmonic relative humidity from Jun 1980 to Oct 2000 Reanalyses (NCEP): a) Variance (%), c) Phase (month), e) Amplitude (%) Simulations (ECHAM5): b) Variance (%), d) Phase (month), f) Amplitude (%)

Chapter 4

4.0 Burn Risk Model: Haines Index (HI)

This chapter discusses atmospheric susceptibility to fires for two climate periods. The current period between 1980-2000 and the simulated future climate based on the IPCC A1B scenario (2080-2100). The analyses are based on the low severity atmospheric index, the Haines index (HI), based on the monthly climatology of the two periods.

For several years, characteristic values of atmospheric stability conditions and water vapor content in the atmosphere have been observed and related to the spread of hotspots (fires) in large remote areas in the United States (BROTAK and REIFSNYDER, 1977). According to Nunes (2005), there are two types of factors that determine the risk of fire. The constant factors: type of combustible material, type of forest and relief. And variable factors, including the humidity of the combustible material and climatic conditions (for example: relative humidity, temperature, rainfall and atmospheric stability).

Climatic conditions and humans maintain a close relationship, from the occurrence of fire, resulting from the atmospheric conditions prevailing in a given period of time, to its maintenance and spread. The Haines index (HI), as noted above, is an indicator for the potential development of convective fire activity as a function of vertical temperature and air humidity profiles calculated in the lower, middle and upper atmosphere (HAINES, 1988). The instability of the air near the fires affects the behavior of the fire, as it increases the vertical dimensions of the smoke columns, resulting in surface winds due to the thermal difference near the burn site. This is the mechanism by which fires create their own winds and the nearby atmosphere becomes unstable. (kocHTUBAJDA, 2001).

Haines (1988) was the first researcher to develop an index based on atmospheric stability conditions and relative humidity at the 925hPa, 850hPa,

700hPa and 500 hPa levels. His results indicated clear evidence of a strong relationship between conditions of atmospheric stability, dry air, low humidity and fire growth. Haines (1988) also concluded that the temperature of the air (T) and the dew point (Td) in different layers of the atmosphere, together with the low level of relative humidity (RH) at the surface, interfere significantly with the development of fires over large areas. The Haines index - HI (HAINES, 1987) depends directly on the level of the atmospheric layer. The combination of three layers in the atmosphere (Lower, Middle and Upper) is therefore used to produce this index, which is the sum of the atmospheric stability terms and the air humidity terms.

This index can also be combined with meteorological parameters such as the previous day's maximum temperature, average accumulated rainfall and wind speed. Recently, HI studies have been carried out for Brazil, in which it was found that the central region of Brazil and the northern strip of northeastern Brazil (NEB) are the regions most prone to the development of fires (LEMOS, 2006).

Mathematically, the calculation of the Haines index (HI) is defined by the sum of the stability component (A) and the atmospheric humidity component (B), as shown in table 1.

IH=A+B

where: the **term A** represents the difference in air temperature between two levels of the atmosphere, and the **term B** characterizes the differences between air temperatures and dew point temperatures for each level of the atmosphere, i.e. the dew point depression

In order to take into account the altitude of the site, the HI is calculated at 3 pressure levels: low altitude, below 1500m (950 hPa- 850 hPa); medium altitude from 1500m to 3500m (850 hPa - 700 hPa); and high altitude (700 hPa - 500 hPa), which corresponds roughly to the vertical profile from the surface to 5000m. **Table 1** shows the entire sequence of calculations.

Altitude	Stability component (A)		Moisture component (B)	
	Calculation	**Category**	**Calculation**	**Category**
low (≤1500m)	A=T950hPa - T850hPa	A=1 if < 4°C A=2 if 4-7°C A=3 if ≥ 8°C	B=T850hPa - Td850hPa	B=1 if < 6 °C B=2 if 6-9 °C B=3 if ≥10 °C
average (1500 3500m)	A=T850hPa - T700hPa	A=1 if < 6°C A=2 if 6-10°C A=3 if ≥11°C	B=T850hPa - Td850hPa	B=1 if < 6 °C B=2 if 6-12°C B=3 if ≥13°C
high (≥3500m)	A=T700hPa - T500hPa	A=1 if < 18°C A=2 if 18-21°C A=3 if ≥ 22°C	B=T700hPa - Td700hPa	B=1 if <15 °C B=2 if 15 20°C B=3 if ≥21 °C

Table 1: Illustration of the different components of the Haines Index (HI) Source: Winkler *et al.* (2005)

The absolute values of the differences are converted into indices ranging from 2 to 6, as shown in **Table 2**, and characterize the risk of fires as very low (2.3), low (4), moderate (5) and high risk (6).

A+B = IH	Risk
2,3	very low
4	low
5	moderate
6	high

Table 2: Calculation of the Haines index

Source: Winkler *et al.* (2005)

In the analyses, it was decided to present the fire risk (HI) according to temperature, as follows: very low index, with values below 12°C; low index with values between 12°C and 20°C; moderate index between 20°C and 24°C and high index above 24°C, as **shown in Table 3.**

Risk	IH
very low	≤12°C
low	12°C-20°C
moderate	20°C-24°C
high	≥24°C

Table 3: Calculation of the index as a function of temperature

4.1 HAINES INDEX ANALYSIS

The following shows the behavior of the A, B and IH terms for the NCEP and ECHAM5 data for the months of June, July, August, September and October, in the middle atmosphere, as this layer best represents the atmospheric conditions favorable to fires (LEMOS, 2006).

It should be noted that most of the fires in Brazil are of anthropogenic origin, even though the risk of fires is maximum, and this does not imply that it is in the presence of fire. The case is different in North America, where most fires are of natural origin. The IH only provides the atmospheric favorability for the occurrence of forest fires.

4.1.1 Analysis for the month of June

Figures (4a, b, c) and Figures (4.1a, b, c) were generated for the month of June from NCEP and ECHAM5 data, respectively. It can be seen that in the middle atmosphere, the stability component (term A) of the Haines index for the NCEP data (Fig. 4a) has a category (2), with temperature values between 6°C-10°C for the whole of Brazil. The humidity component (term B) (Fig. 4b) has a category (1), with temperatures below 6°C in the central-western region and western Minas Gerais. Based on the sum of the A and B terms (A + B = IH), there are temperature values below 12°C (low risk) for the Brazilian coast, and values between 12°C-20°C in most of the central-western region of Brazil (Fig.

4c), which characterizes a moderate risk.

The analysis of the ECHAM5 data shows that there is greater instability (term A) in eastern Parà and western Amazonia (Fig. 4.1a), when compared to the NCEP results. The humidity component (term B) shows higher relative humidity (Fig. 4.1b), with differences between air temperature and dew point of less than 6°C for the whole of Brazil. Thus, the sum of the values shows that for the month of June, according to the HI, the risk of major fires is small (Fig. 4.1c).

In order to investigate the atmospheric favorability to fires at the end of the 21st century, Figures (4.2a, b, c) were generated for future simulations (2080-2100) using the ECHAM5 model. Term A (Fig. 4.2a) shows category (3) in the northern region and the state of Mato Grosso, and category (2) for the other regions. Term B (Fig. 4.2b), on the other hand, has category (1) in most of Brazil, since the dew point depression is less than 6°C. Based on these analyses, no significant changes in atmospheric susceptibility to wildfires should be expected by the end of the century. Although there is an area of moderate risk in the western part of Mato Grosso based on the analysis of the data projected for the future.

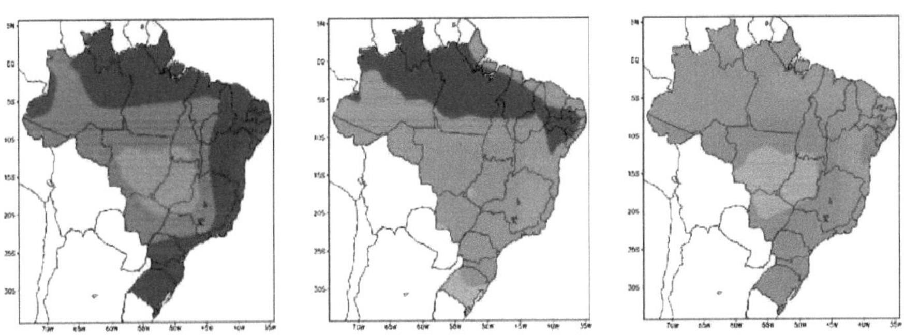

a) b) c)

Figure 4: a) Term A, b) Term B, c) Haines index (HI), (M-P-NCEP-JUN/6)

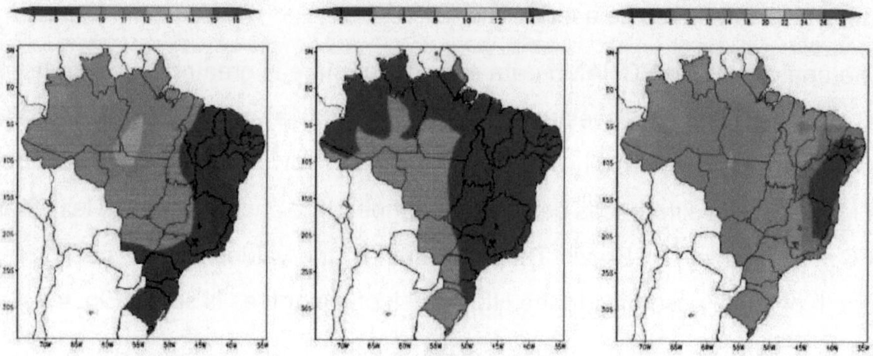

a) b) c)

Figure 4.1: a) Term A, b) Term B, c) IH, (M-P-ECHAM-JUN/6)

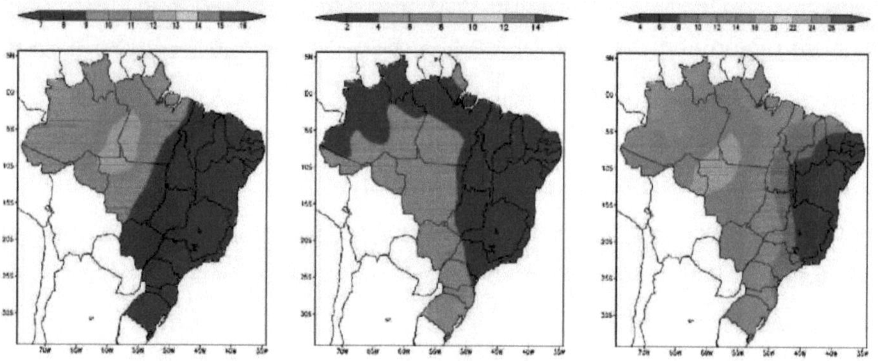

a) b) c)

Figure 4.2: a) Term A, b) Term B, c) IH, (M-AG-ECHAM-JUN/6)

In the interests of a more detailed investigation, we decided to present Figures 4.3a, b with the fires detected by the NOAA-12 satellite for the month of June in 2006 and 2007.

In June, the fires detected by the NOAA-12 satellite reached approximately 3,150 outbreaks for 2006 (Fig. 4.3a) and around 4,900 outbreaks for 2007 (Fig. 4.3b). The areas with the highest occurrence of fires were the southeast and central-west regions of Brazil. A brief assessment shows that these areas also have the highest HI values (Figs. 4c, 4.1c).

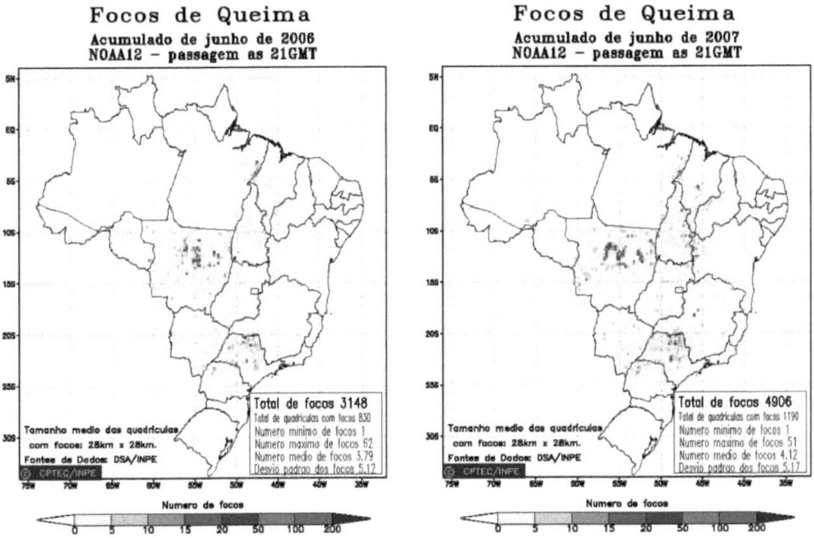

a) b)

Figura 1.3: **a) Accumulated fires in June 2006**

b) Accumulated fires in June 2007

Source: Center for Weather Forecasting and Climate Studies

4.1.2 Analysis for July

The analyses for July, according to the NCEP data, show that in the middle atmosphere term A (Fig. 4.4a) has a category (2) for the north, northeast, south, southeast and Mato Grosso regions, and a category (3) for the states of Mato Grasso do Sul and Tocantins. Term B (Fig. 4.4b) shows category (1) for the north, northeast and southeast regions and in the states of Paranà and Santa Catarina; and category (2) for the central-west region and the state of Rio Grande do Sul. As a result, the Haines index (HI) (Fig. 4.4c) is characterized as low (4) in practically all of Brazil, except for central Mato Grosso and eastern Goiás, where it is moderate (5).

The comparison between the ECHAM5 data and the NCEP data shows an intensification of the magnitude of both the values of the stability component (term A) (Fig. 4.5a), implying a more unstable atmosphere, and the humidity term (term B) (Fig. 4.5b), which suggests a drier atmosphere. Generally speaking, this intensification of the A and B terms does not result in major changes in the final HI value compared to the NCEP results. It should be noted that the north of Mato Grosso and the south of Parà present a moderate risk of fires.

Comparing the months of June and July, there is a clear intensification of the HI values, as well as an increase in the area of maximum risk for the month of July. The HI analyses for the future global warming scenario clearly show an intensification of the A and B terms (Fig. 4.6a, b, c). This shows that the whole area, from the north of Mato Grosso, west of Parà and east of Amazonas, will be very vulnerable to the development of large-scale fires if the climate forecasts are confirmed.

In addition, the A term, which reproduces atmospheric stability or convective power, dominates the process, leaving a smaller contribution to the B term, which illustrates the water vapor content at 850 hPa. This intensification of HI in July compared to June for both seasons is well accompanied by an increase in fire outbreaks (Fig. 4.7a, b). In other words, despite the differences between the areas of maximum risk and the location of the hotspots, there is a good relationship between the two, particularly for the HI of the NCEP data and the hotspots.

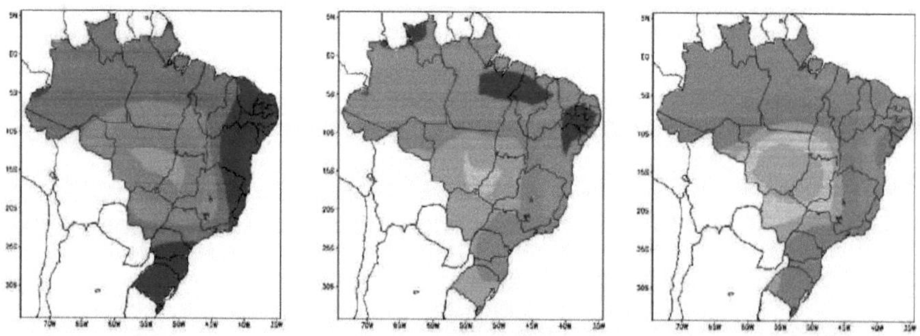

a) b) c)

Figure 4.4: a) Term A, b) Term B, c) Haines index (HI), (M-P-NCEP-JUL/7)

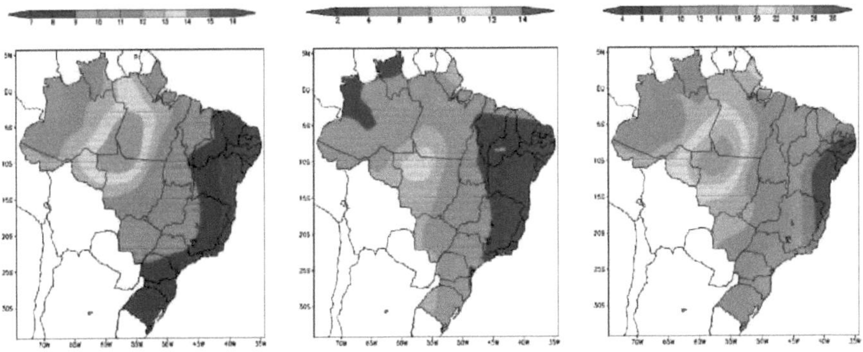

a) b) c)

Figure 4.5: a) Term A, b) Term B, c) IH, (M-P-ECHAM-JUL/7)

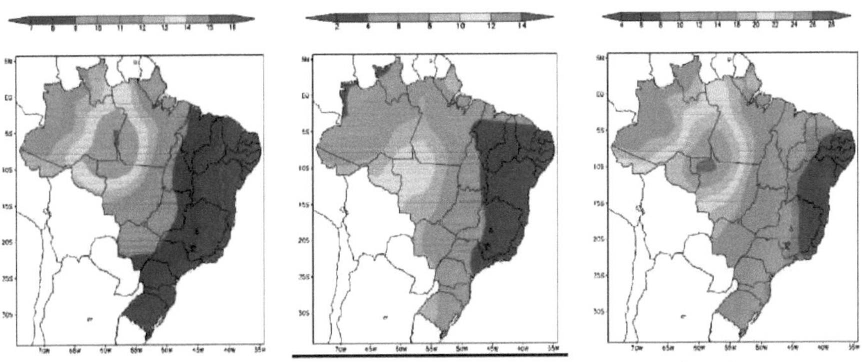

a) b c)

Figure 4.6: a) Term A, b) Term B, c) IH, (M-AG-ECHAM-JUL/7)

The fires reflect, above all, a human practice that follows the natural seasonality, following the occurrence of drought in most of the country. According to data from the NOAA-12 satellite, in July 2007, the number of hotspots detected in Brazil was higher than in the same period in 2006: 6,973 (Fig. 4.7a) compared to 7,056 (Fig. 4.7b).

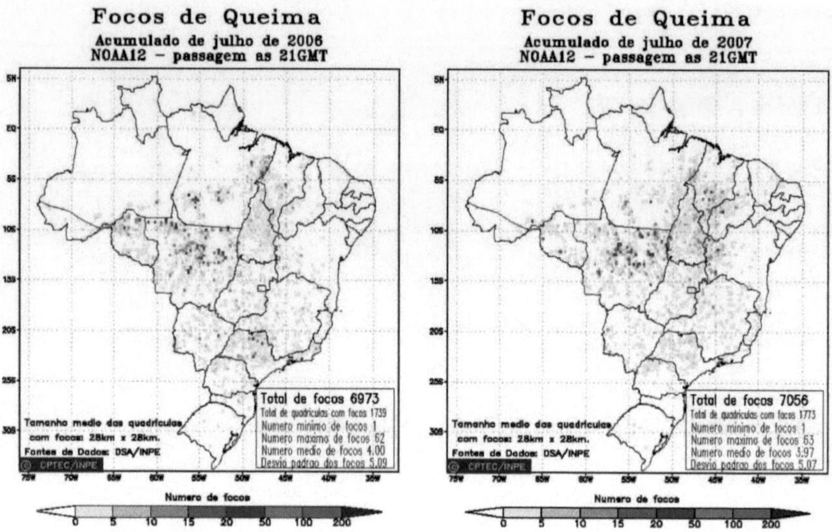

a) b)

Figure 4.7: a) Accumulated fires in July 2006

b) Accumulated Fires in July 2007

Source: Center for Weather Forecasting and Climate Studies (CPTEC)

4.1.3 Analysis for August

In the mid-atmosphere for the month of August, the stability component (term A) for the NCEP data (Fig. 4.9a) has a category (2) for the south, northeast and north of the northern region. The southeast of Amazonas, the south of Parà and the central-west region have a category (3). The humidity component (term B) (Fig. 4.9b) has a category (1) in the northern part of the northern region and in the northeastern region; a category (2) for the southern, southeastern and central-western regions, and higher values (3) for central-southern Mato Grosso. As a result, the HI (Fig. 4.9c) is characterized by high risk for the central-western region with temperatures above 24°C, and moderate risk (5) for the other regions.

The ECHAM5 results, when compared to the NCEP results, show a larger area covered by term A with a different spatial distribution of category (3), which suggests greater atmospheric instability (Fig. 4.10a). For term B, the model shows a wetter climate than that proposed by the NCEP, since the difference between the air temperature and the dew point temperature is smaller for ECHAM5 (Fig. 4.10b). However, this humidity is not enough to reduce the risk of fire in the area stretching from Mato Grosso to Pará (Fig. 4.10c).

Investigating the conditions proposed by the global warming scenario (Fig. 4.11 a, b, c), we see a substantial intensification of the conditions favorable to the development of fire. This is clearly seen in the states of Mato Grosso, Para and Amazonas (Fig. 4.11c). The evaluation of the previous months (June, July) illustrates that with the start of the dry season in central Brazil, the atmospheric possibility/favorability for the occurrence of fires increases. Figure 4.8a, b shows the distribution of fire outbreaks for the month of August in 2006 and 2007, according to the NOAA-12 satellite. In fact, the north-central region of Brazil has the highest number of fires.

In 2006, around 25,800 fires were detected, while in 2007, around 21,450 were detected.

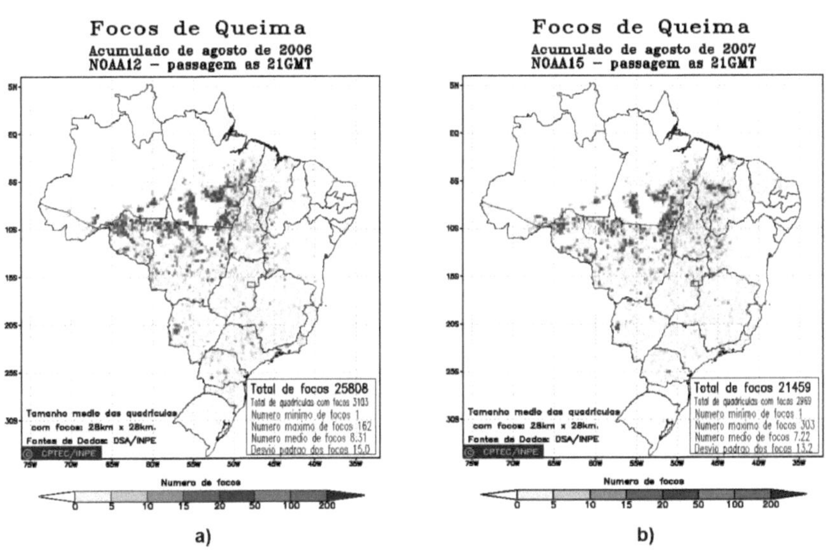

Figure 4.8: a) Accumulated fires in August 2006. b) Accumulated fires in August 2007.

Source: Center for Weather Forecasting and Climate Studies (CPTEC)

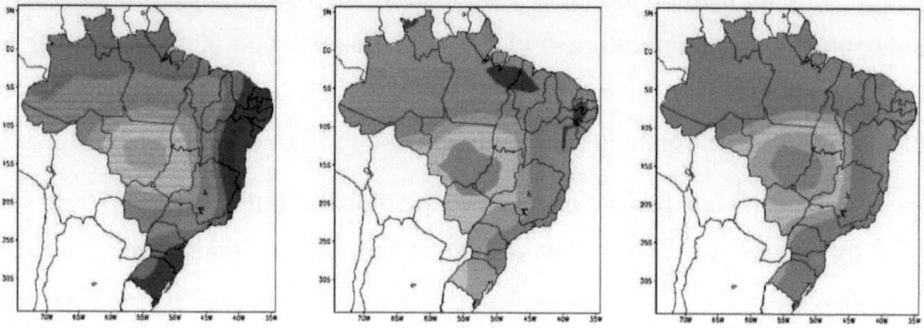

a) b) c)

Figure 4.9: a) Term A, b) Term B, c) Haines Index (HI), (M-P-NCEP-AGO/8)

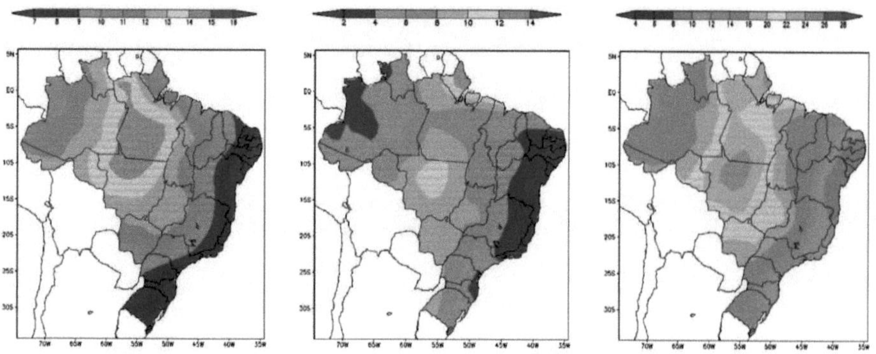

a) b) c)

Figure 4.10: a) Term A, b) Term B, c) IH, (M-P-ECHAM-AGO/8)

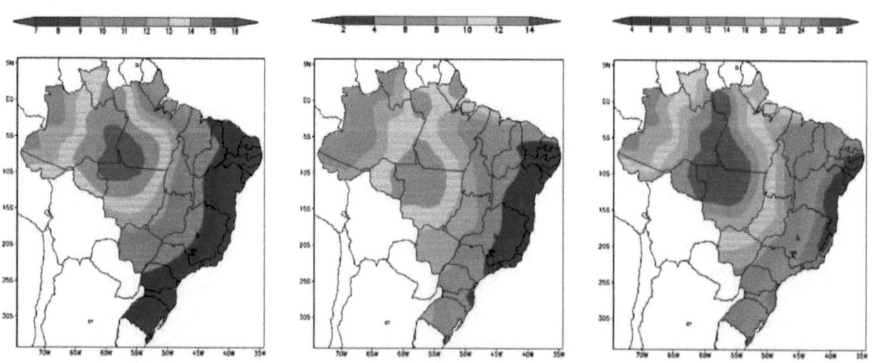

a) b) c)

Figure 4.11: a) Term A, b) Term B, c) IH, (M-AG-ECHAM-AGO/8)

4.1.4 Analysis for the month of September

According to NCEP data, during September there is a decrease in the intensity of the A and B terms and the Haines index compared to August. Although the risk area in central Brazil can still be observed. The analysis of the ECHAM5 results (Fig. 4.13a, b, c) is totally different in terms of spatialization and magnitude when compared to the NCEP data (Fig. 4.12a, b, c). While the NCEP results show the central region of Brazil as more unstable and drier, the ECHAM5 results show Parà and the west of northeastern Brazil (NEB) as more favorable to fires.

For a future global warming scenario, the HI shows a notable intensification of the A term (Fig. 4.14a). Therefore, convective power is responsible for a large part of the process that results in the risk of fires, while the water vapor content in the atmosphere at 850 hPa has a smaller contribution. This suggests the possibility of large areas with greater susceptibility to burning, namely the central-western and northern regions. This reduction in HI in September compared to August is not in line with the increase in the number of hotspots as shown in Figures 4.15a, b and 4.8a, b. Areas with a high rate of fires, such as Rondônia and eastern Para, do not show high risk conditions according to the HI.

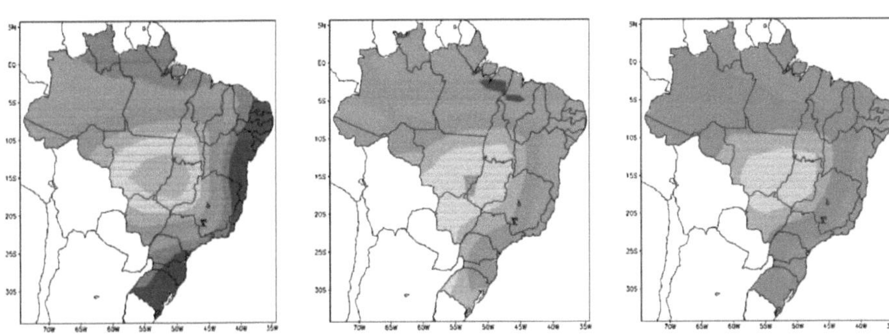
a) b) c)

Figure 4.12: a) Term A, b) Term B, c) Haines Index (HI), (M-P-NCEP-SET/9)

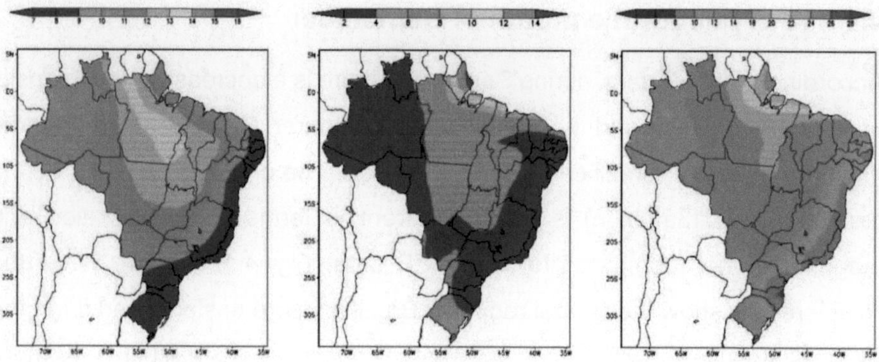

a) b) c)

Figure 4.13: a) Term A, b) Term B, c) IH, (M-P-ECHAM-SET/9)

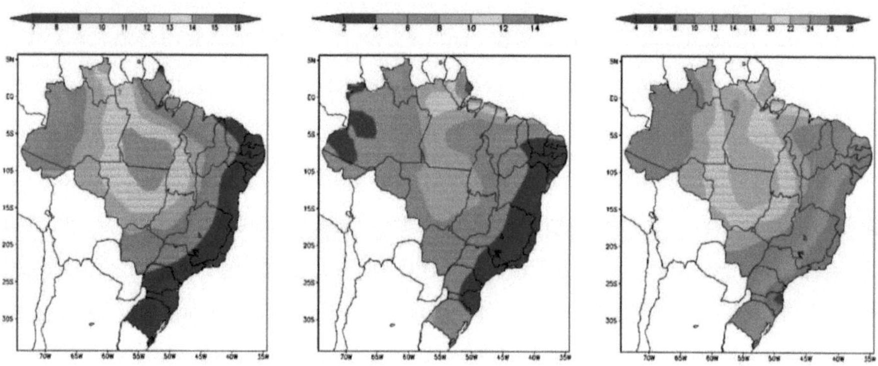

a) b) c)

Figure 4.14: a) Term A, b) Term B, c) IH, (M-AG-ECHAM-SET/9)

This number corresponds to the month with the highest incidence of fires, especially in the states of Parà, Tocantins and northern Mato Grosso. Therefore, the most reliable results are those from ECHAM5, as they are more consistent in terms of the number of fires detected.

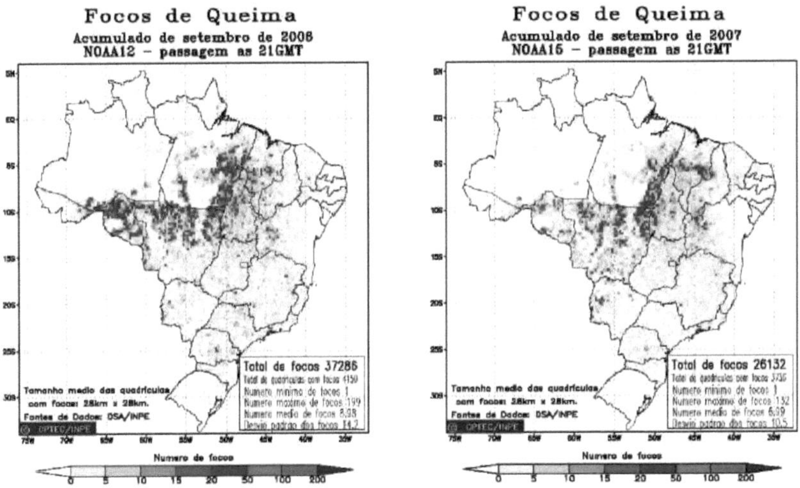

b) a)

Figure 4.15: a) Accumulated fires in September 2006

b) Accumulated fires in September 2007

Source: Center for Weather Forecasting and Climate Studies (CPTEC)

4.1.5 Analysis for the month of October

For the middle atmosphere, the A term for the NCEP data (Fig. 4.16a) shows a category (3) in the central region of Brazil, and part of the southeast, with corresponding maximum temperatures of 13°C. The humidity component (term B) (Fig. 4.16b) has a category (2) below 10°S and a category (1) above 10°S. As a result, the HI is shown to be a low risk (4) for the whole of Brazil, with temperatures below 18°C.

For the ECHAM5 results (Fig. 4.17a, b, c), the stability component is more stable, and the dew point depression, the B term, is wetter, implying that the HI is at very low risk (2.4) and low risk (4). As such, this month is not very favorable for burning. Similar to what was described above, the future simulations (2080-2100) (Fig. 4.18a, b, c), are characterized by a low risk (4) for the whole of Brazil with values below 16°C.

This is not matched by the heat spots (Fig. 4.19a, b). According to this figure, there is still a high rate of fires in Brazil, especially in Para and Maranhão.

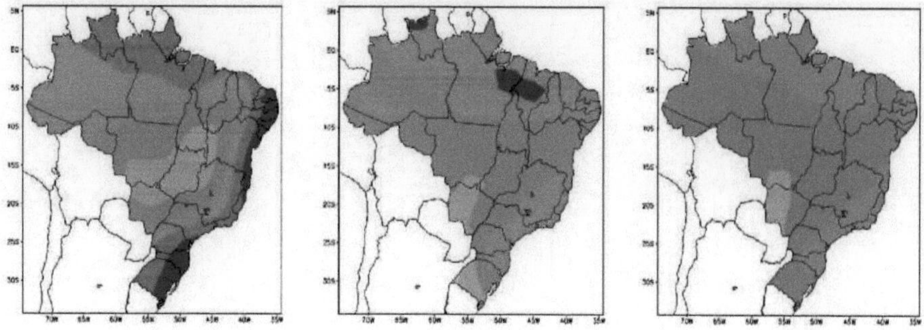

a) b) c)

Figure 4.16: a) Term A, b) Term B, c) Haines index (HI), (M-P-NCEP-OUT/10)

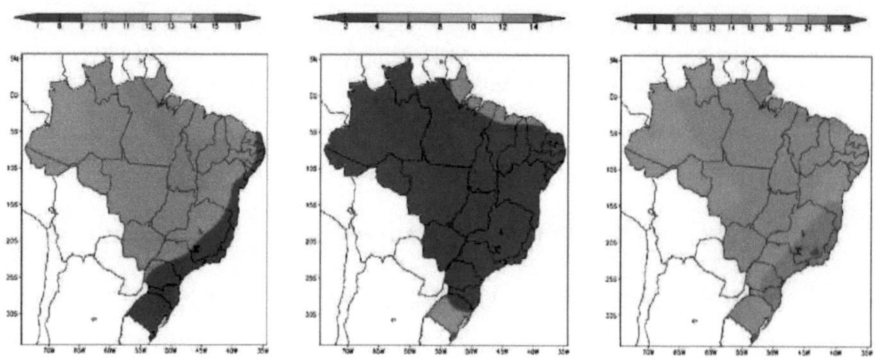

a) b) c)

Figure 4.17: a) Term A, b) Term B, c) IH, (M-P-ECHAM-OUT/10)

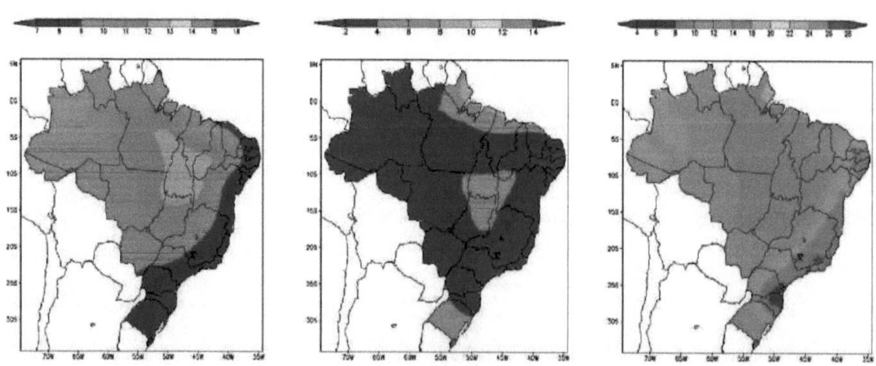

a) b) c)

Figure 4.18: a) Term A, b) Term B, c) IH, (M-AG-ECHAM-OUT/10)

It is important to note that these results cannot be understood solely as a flaw in the HI methodology, but rather as a possible limitation in the representation of the climate as proposed by the NCEP, as well as the climate modeled by ECHAM5. In addition, the analyses shown are based on monthly averages, which can smooth the distribution of the A, B and IH terms. Calculating on a daily basis may provide more satisfactory results. Another point to note is the low spatial resolution of the data. Lemos (2006), using daily data from the ETA model, found a good correlation between the outbreaks detected and the areas of maximum risk.

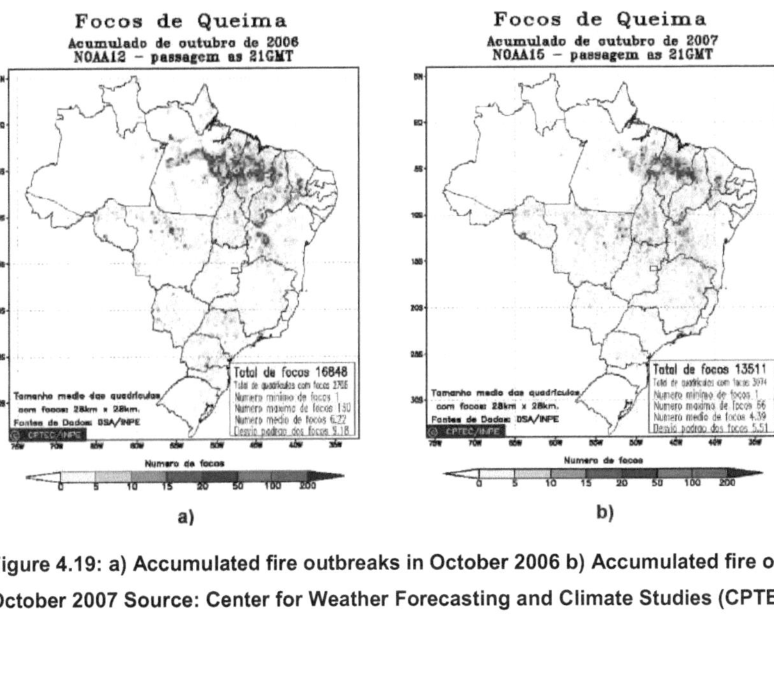

a) b)

Figure 4.19: a) Accumulated fire outbreaks in October 2006 b) Accumulated fire outbreaks in October 2007 Source: Center for Weather Forecasting and Climate Studies (CPTEC)

Chapter 5

5.0 Burn Risk Model: Setzer Index

5.1 Fire risk concept.

This Fire Risk (RF) methodology was developed internally at CPTEC, based on an analysis of the occurrence of hundreds of thousands of fires in the main biomes (vegetation types) of Brazil, depending on the meteorological conditions and history in the area of each event (SETZER *et al.,* 2002; SISMANOGLU *et al.*, 2002). Its principle is that the more days without rain, the greater the risk of vegetation burning; additionally, the type and natural cycle of vegetation defoliation, maximum temperature and minimum relative humidity, as well as the presence of fire in the region of interest are included in the calculation. The reference for the calculations is the "Days of Drought", or "Secura", (S), which is a hypothetical number of days without any precipitation during the last 120 days (SETZER and SISMANOGLU, 2006).

5.1.1 Sequence of the observed fire risk calculation

Below are the steps for obtaining the fire risk, proposed by Setzer *et al.* (2002) and defined in this work as the Setzer index (SI).

1) Determines the daily accumulated rainfall, in millimeters (*mm*), for the geographic area covered, for eleven immediately preceding periods, from 1; 2; 3; 4; 5; 6 to 10; 11 to 15; 16 to 30; 31 to 60; 61 to 90 and 91 to 120 days.

2) Calculates the "Precipitation Factors" (PF), with values from 0 to 1, for each of the eleven periods, by means of an empirical exponential function of the precipitation in millimeters of rain for each of them. The equations are respectively:

$$FP_1 = \exp(-0{,}14prec)\ ; \quad FP_2 = \exp(-0{,}07prec)\ ;$$

$$FP_3 = \exp(-0{,}04prec)\ ; \quad FP_4 = \exp(-0{,}03prec);$$

$$FP_5 = \exp(-0{,}02prec)\ ; \quad FP_{6\,a\,10} = \exp(-0{,}01prec);$$

$$FP_{11a\,15} = \exp(-0{,}008prec)\ ; \quad FP_{16\,a\,30} = \exp(-0{,}004prec);$$

$$FP_{31a\,60} = \exp(-0{,}002prec)\ ; \quad FP_{61a\,90} = \exp(-0{,}001prec);$$

$$FP_{91a\,120} = = \exp(-0{,}0007prec)$$

3) Calculate the "Dryness Days" (DW) by multiplying the FP according to the equation:

$$PSE = 105.FP_1.FP_2 \ldots FP_{91a120}$$

4) Determine the basic fire risk (RB) for each of the five types of vegetation considered, using the equation:

$$RB_{n=1,5} = 0{,}9[1 + sen\,(A_{n=1,5}.PSE)]/2$$

CLASS OF VEGETATION	**1**	**2**	**3**	**4**	**5**
TYPE OF VEGETATION	Dense Ombrophylous	Ombrophilous Open	Contact Campinarama	Seasonal+ +Deciduous+ Semi-Decide	No Forest
CONSTANT "'A"	1,715	2	2,4	3	4,7

Table 4: Vegetation class used in IS

Source: Setzer *et al.*, 2002

The Basic Risk (RB) follows a sinusoidal curve over time, since global solar radiation and the vegetation's response show roughly the same pattern. The risk therefore increases according to this curve, reaching a maximum value of 0.9. Figure 5 illustrates the variation in basic risk using the above equations and concepts. It can be seen that the "days without rain" axis indicates both a real period of days without rain, as well as "dry days", which correspond to a hypothetical period without rain calculated from the quantity and temporal distribution of rainfall. The notion of non-rainy periods - "days without rain" or "dry days" - is the basic principle of this method.

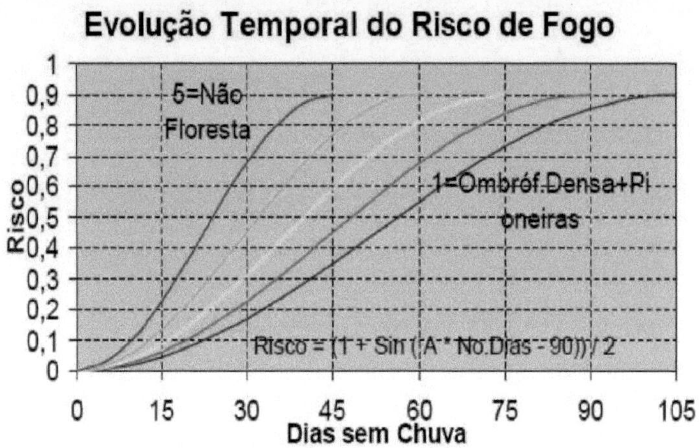

Figure 5: Variation in fire risk (RF) for different vegetations Source: Center for Weather Forecasting and Climate Studies (CPTEC)

In addition to the dryness factor, there are two other factors for the risk of final fire.

1) Corrects the risk of fire for the minimum relative humidity. The risk (RU) increases for humidity below 40% and decreases for values above this reference. Humidity data from 6pm UTC observations is used, supposedly its minimum. The linear fit equation is:

$$RU = RB(-0,006.UR_{min} + 1,3)$$

2) Corrects the risk of fire for the maximum air temperature. The risk (RT) increases for temperatures above 30°C and decreases for values below this reference. Humidity data from observations at 18:00 UTC, supposedly its maximum, is used. The linear adjustment equation is:

$$RT = RU(0{,}02T_{máx} + 0{,}4)$$

$$Risco\ de\ Fogo\ Final(RF) = RT$$

Once the calculations have been made, the risk as a function of the maximum temperature is the final fire risk (RF), and the following categories are assigned:

RISK LEVEL	SETZER INDEX (0 to 1)
minimum	0-0.15
low	0.15 -0.4
medium	0.40 -0.7
high	0.70 - 0.90
critical	> 0.90

Table 5. Fire risk classes

The tool used to perform the calculations was the Fortran programming language in a Linux environment. The maximum temperature (Tmax), minimum relative humidity (URmin) and rainfall data were extracted from the model outputs on a daily basis.

The second phase of the process consisted of applying the risk equations to the input variables Tmax, RHmin and precipitation, still on a daily basis; this amounted to approximately 43800 files for the two eras: current and global warming. Due to the difficulty in showing the Setzer index (SI) on a daily basis, we opted to present it from the monthly climatology with daily data.

5.2 CHANGE IN VEGETATION

For decades, the scientific community has been intrigued by the challenge of research into global change, particularly in predicting the future distribution of vegetation. This question of the possible occurrence of vegetation change in a given region and its impact on the climate, whether due to natural or anthropogenic forest action, can have a major impact on the climate, not only locally, but also regionally and globally (CHARNEY *et al.*, 1977; SUD AND FENESSY, 1982 and 1984; XUE AND SHUKLA, 1993 and 1996; DIRMEYER AND SHUKLA, 1996; OYAMA, 2002; COOK AND VIZY, 2008). An assessment of the climatic consequences of changes to natural vegetation includes, among other factors, an understanding of atmospheric and surface conditions. For example, if the "savannization" of the Amazon were to occur, what would this impact imply in relation to the risk of fire in the Amazon forest?

Recent studies such as those by Cook and Vizy (2008), based on a regional climate model coupled with a potential vegetation model, propose a reduction of around 70% of the Amazon rainforest (dense ombrophilous) to cerrado (semi-deciduous) by the end of the 21st century. This current and future distribution of the Amazon rainforest and other biomes (COOK AND VIZY, 2008) is shown in Figures 5.1a, b. As the Setzer index uses vegetation as its input function, this index is calculated assuming these two vegetation scenarios.

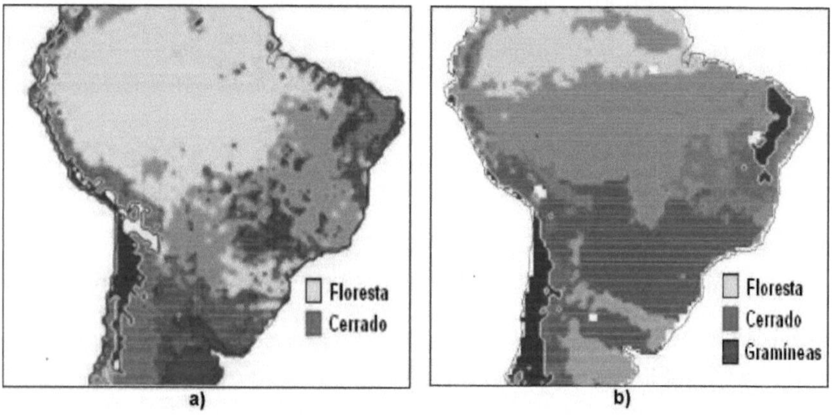

Figure 5.1: a) Current vegetation distribution in South America, b) Predicted future vegetation distribution in South America. Source: Cook and Vizy (2008).

Three assessments are made: the first investigates the risk of fire based on the current climate and vegetation (Fig. 5.1a). The second analysis is based on the future climate (2080-2100) and does not take into account changes in vegetation, using the distribution of biomes as at present. The third calculation is made using the simulations proposed for the future climate and vegetation as for the future scenario, in which the forest vegetation (dense ombrophilous) changes to savannah (semi-deciduous), and areas of savannah become grass (not forest) as shown in Figure 5.1b.

5.3 SETZER INDEX (SI) ANALYSIS

Similar to the analysis of the Haines index, what follows is an investigation of the atmospheric favorability to the occurrence of wildfires. This analysis is based on the risk factors that make up the Setzer index (IS): the maximum temperature factor (Tmax), the humidity factor (URmin) and the dryness index (PSE), which is a function of the type of vegetation and the accumulated rainfall over 11 periods.

5.3.1 Analysis for the month of June

First, an assessment of the risk of fires for the month of June is made based on the current and future climate as simulated by ECHAM5, based on the IPCC's A1B scenario. In this first analysis, it is assumed that the vegetation at the end of the century will be the same as in the current period. The analysis is based on the differences between the future and current climate, using maximum temperature, minimum relative humidity and the dry period (PSE) as variables. These variables are therefore the basis of the Setzer Index (SII). The Tmax anomalies clearly show a warming of around 10°C (Fig. 5.2a) for the north-central region of Brazil, which is accompanied by a drop in RH under global warming conditions compared to the current period (Fig. 5.2b). It can also be seen that the precipitation index is intensifying, indicating a longer period of drought (Fig. 5.2c). Obviously these anomalous climatic conditions favor higher fire risk conditions when compared to the values under current conditions. This can be seen in Figure 5.3b. Under future conditions of global warming, the states of Mato Grosso, Pará and Bahia will be even more prone to large-scale fires.

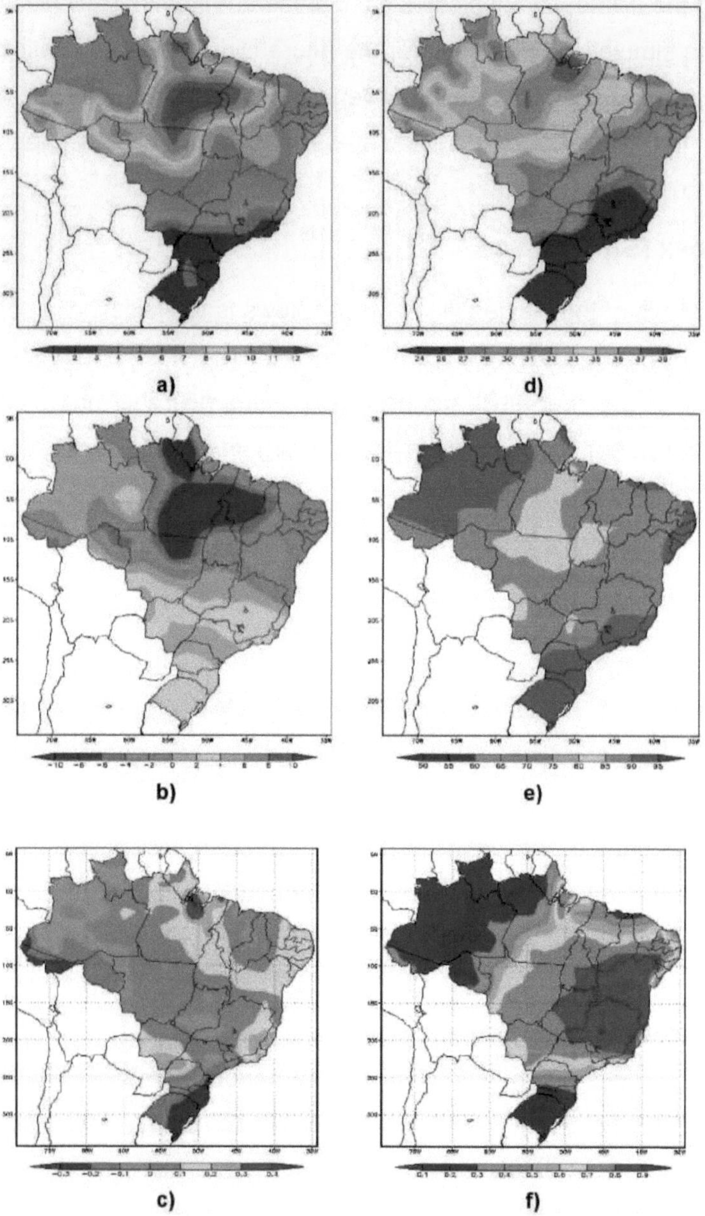

Figure 5.2: a) Tmax anomaly between future climate and current vegetation simulations and the present for June; b) the same as (a) for URmin and c) for PSE, d), e) ,f) shows the spatial distribution of Tmax, URmin and PSE with future climate and vegetation simulations for June.

Figure 5.3c shows a similar analysis. However, in this case we use the vegetation data as predicted by Cook and Vizy (2008) shown in Figure (5.1b). Due to the absence of Amazonian vegetation (forest), there is an intensification of the PSE (Fig. 5.2f) which is a function of the type of vegetation. Therefore, the presence of cerrado induces an increase in PSE and, subsequently, a redefinition of IS or fire risk. It is also noteworthy that the region from Bahia to the north of Minas Gerais has a low risk of fires (Fig 5.3e). Under future conditions, this area will be covered by grasses and shrubs, whereas under current conditions the cerrado and caatinga are the predominant biomes.

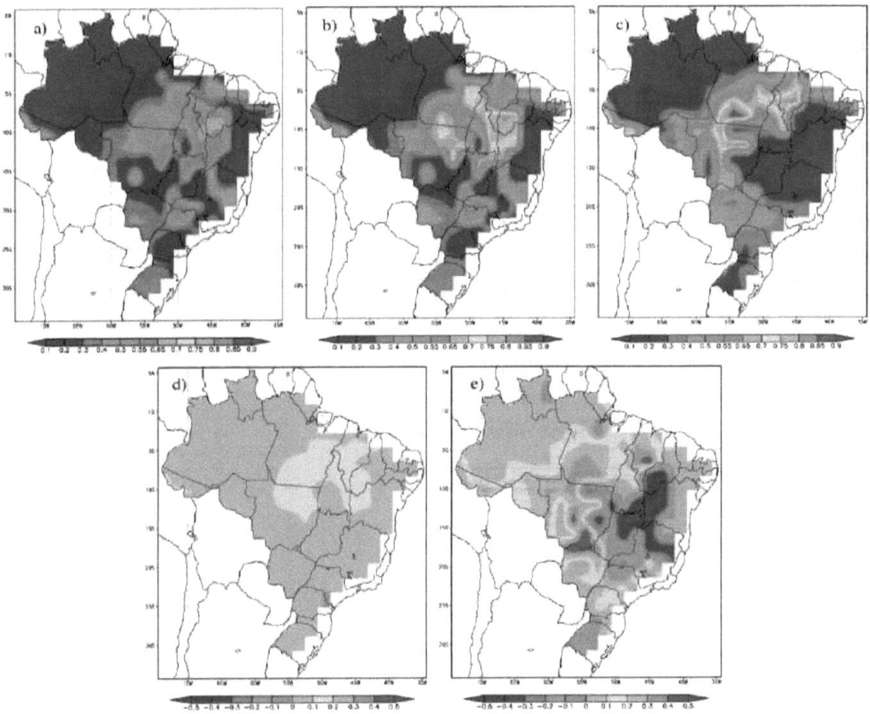

Figure 5.3: a) Setzer index (SI) for current climate and vegetation in June; b) SI for future climate and current vegetation in June; c) SI for future climate and vegetation simulations in June; d) are the anomalies between b) and a); e) are the anomalies between c) and a).

5.3.2 Analysis for July

During the month of July there is an intensification of conditions susceptible to fires compared to June (Fig. 5.4), although the temperature anomalies between the future and current periods are smaller (Fig. 5.4a and 5.2a).

The calculation of the IS (fire risk) shows a greater coverage of the area with the highest risk of fires, mainly in the north of Mato Grosso for the current vegetation conditions and future climate (Fig. 5.5a). However, compared to June, July shows a lower IS in western Bahia and southern Piaui. This must be related to the lower temperature and PSE anomalies (Fig 5.4a, c). When future vegetation is included (Fig. 5.5c), the highest IS values predominate in the state of Rondônia. When compared to the IS calculated for the present (Fig. 5.5a), which can be better visualized by the field of anomalies (Fig. 5.5c), it can be seen that in much of central Brazil the exchange of current vegetation for future vegetation induces a reduction in the risk of fire. On the other hand, in the western part of Brazil (Paranâ, Mato Grosso do Sul, Mato Grosso and Roraima) we can expect an increase in the risk of fires.

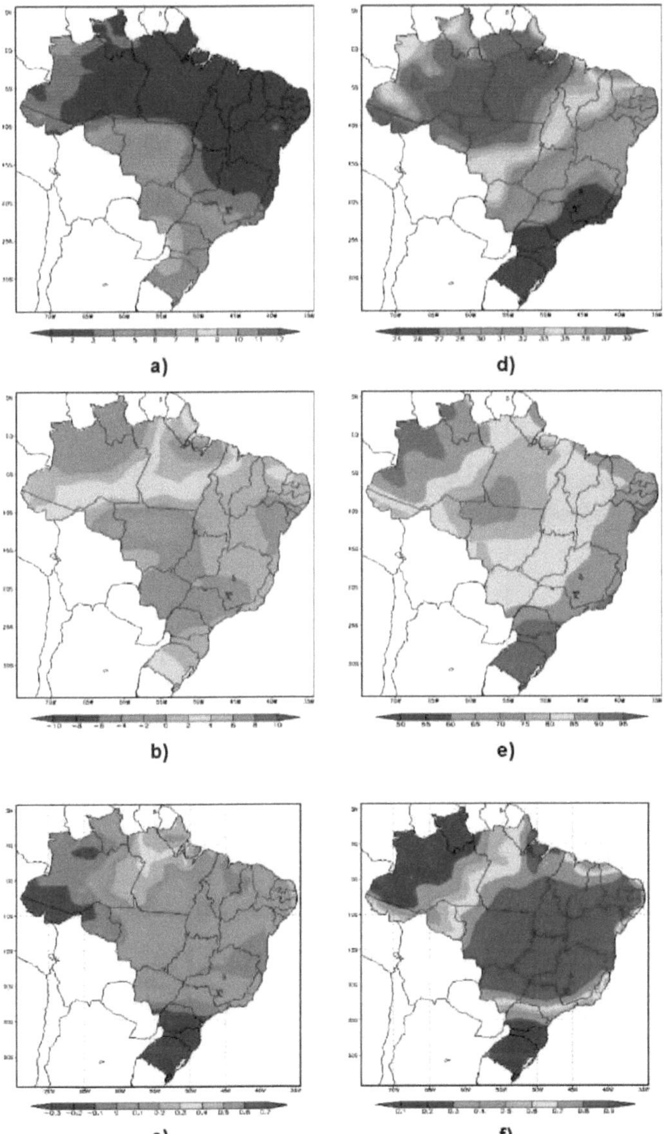

Figure 5.4: a) Tmax anomaly between the simulation of future climate and current vegetation and the present for July; b) the same as (a) for URmin and c) for PSE, d), e) ,f) shows the spatial distribution of Tmax, URmin and PSE with future climate and vegetation simulations for July.

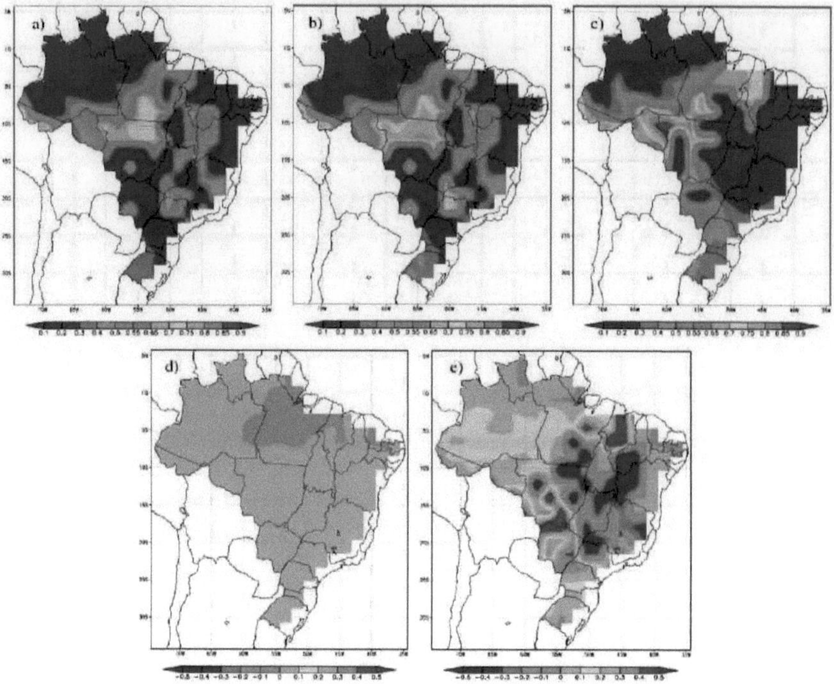

Figure 5.5: a) IS for current climate and vegetation in July; b) IS for future climate and current vegetation in July; c) IS for climate and vegetation simulations in July; d) are the anomalies between b) and a); e) are the anomalies between c) and a).

5.3.3 Analysis for August

According to Figure (4.8a, b), which shows the distribution of fires in August for the years 2006 and 2007, it can be seen that the north of Mato Grosso, Rondônia and the south/east of Parà are the areas with the highest prevalence of fires. In fact, it is in these regions that the highest Tmax and lowest RHmin values are found, as well as the most intense dry period (PSE) (Fig. 5.6a, b,c).

The evaluation of the IS (Fig. 5.7a), calculated on the basis of current climatic conditions, points to the existence of a critical risk in the states mentioned above. The analysis for global warming conditions shows an even greater intensification of the critical zone for the use of fire (Fig. 5.7b). Including the vegetation predicted for the last decades of the 21st century shows a migration of the area most favorable to burning. Under these conditions, the maximum IS now covers the northern part of Parà, eastern Amazonas and Rondônia. It is important to note that these areas, particularly the northwest of Parà, do not have a large number of fires under current conditions, as it is still an area of closed forest with little human presence. Figure (5.7e) shows the anomalies between the current risk and the proposed risk based on climate and future vegetation. There is a considerable increase in the risk of fire, especially in the state of Amazonas and northwest Pará. This is due to the conversion of native forest into cerrado, which is the vegetation proposed by the future scenario. On the other hand, the presence of grasses, replacing the cerrado, reduces IS in Mato Grosso and southern Parà.

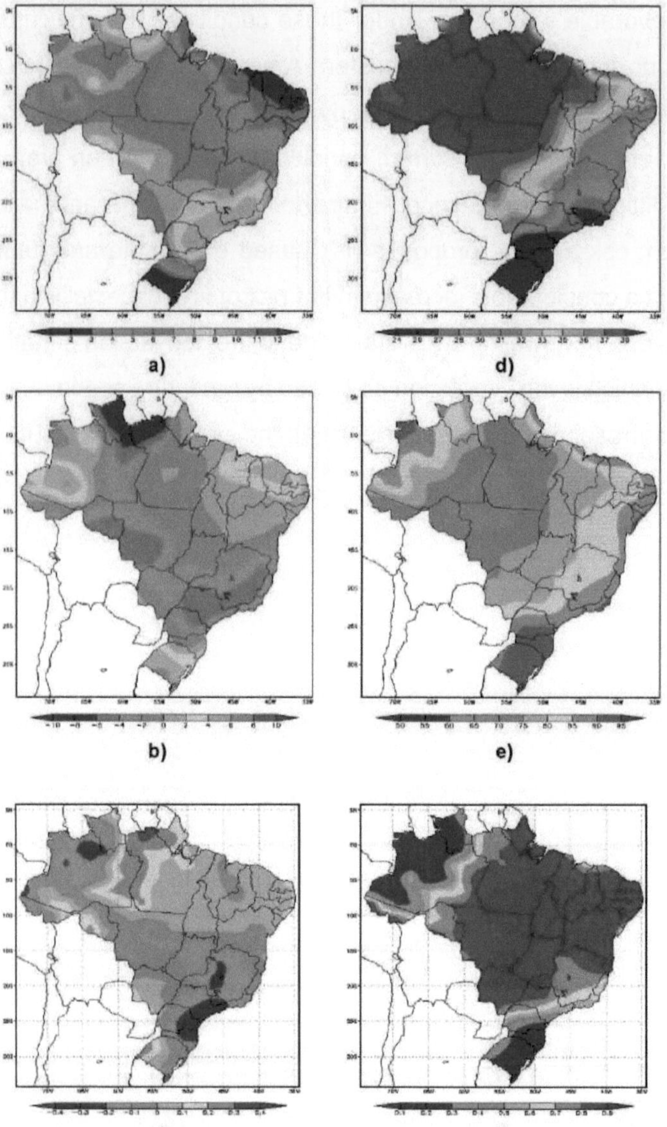

Figure 5.6: a) Tmax anomaly between the simulation of future climate and current vegetation and the present for August; b) the same as (a) for URmin and c) for PSE, d), e) ,f) shows the spatial distribution of Tmax, URmin and PSE with future climate and vegetation simulations for August.

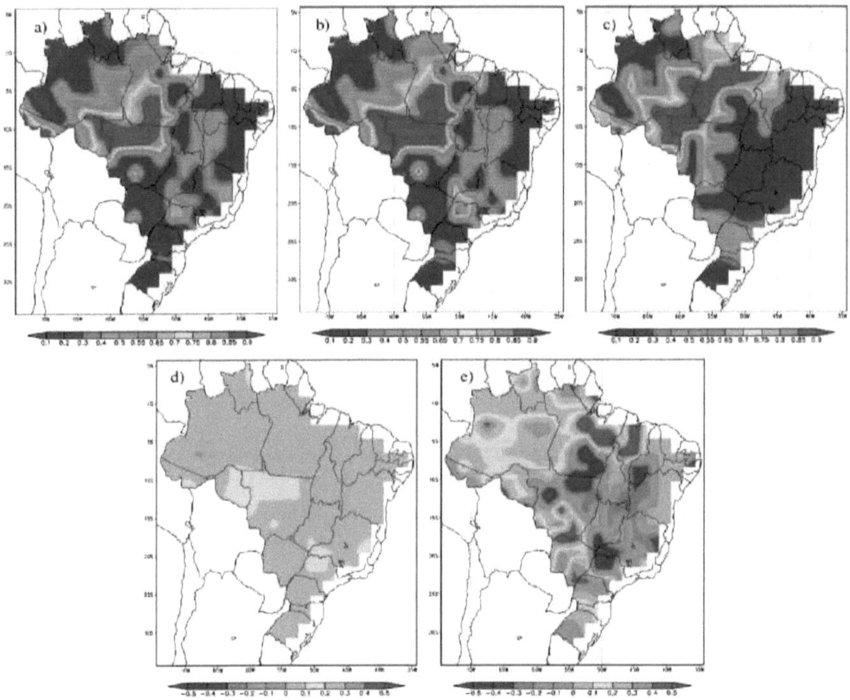

Figure 5.7: a) Setzer index (SI) for current climate and vegetation in August, b) SI for future climate and current vegetation in August, c) SI for future climate-vegetation simulations in August; d) are the anomalies between b) and a); e) are the anomalies between c) and a).

5.3.4 Analysis for September

The evolution of climatic conditions throughout the year leads to considerable changes in the patterns/distributions of the Tmax, URmin and PSE fields. In September, the dominant condition for establishing IS is the distribution of precipitation, or PSE (Fig. 5.8c). It is important to note that the conditions associated with Tmax and RHmin, particularly in the eastern Amazon, are also important.

The central-southern region of Brazil has a unique characteristic in that the highest Tmax anomalies (Fig. 5.8a) and minimum RHmin anomalies (Fig. 5.8b) are not enough to produce a region of critical IS. The fire risk calculated for current conditions (Fig. 5.9a) confirms the areas of maximum risk in Mato

Grosso and Parà, already showing an IS that is characterized by a moderate level. Using climate data predicted for the future (Fig.5.9b), the SI remains over the same areas, but with higher values, characterizing the critical level. For the month of September, the use of the predicted vegetation causes a reduction in SI, except in the extreme north of Parà (Fig. 5.9c).

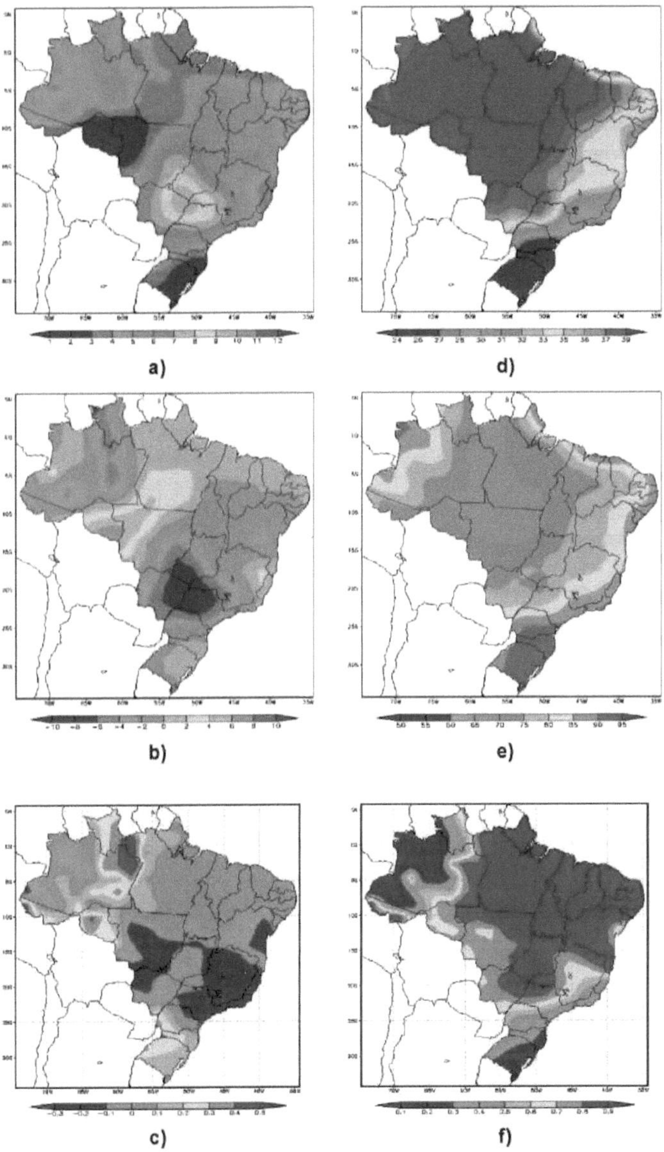

Figure 5.8: a) Tmax anomaly between the simulation of future climate and current vegetation and the present for September; b) the same as (a) for URmin and c) for PSE, d), e) ,f) shows the spatial distribution of Tmax, URmin and PSE with future climate and vegetation simulations for September.

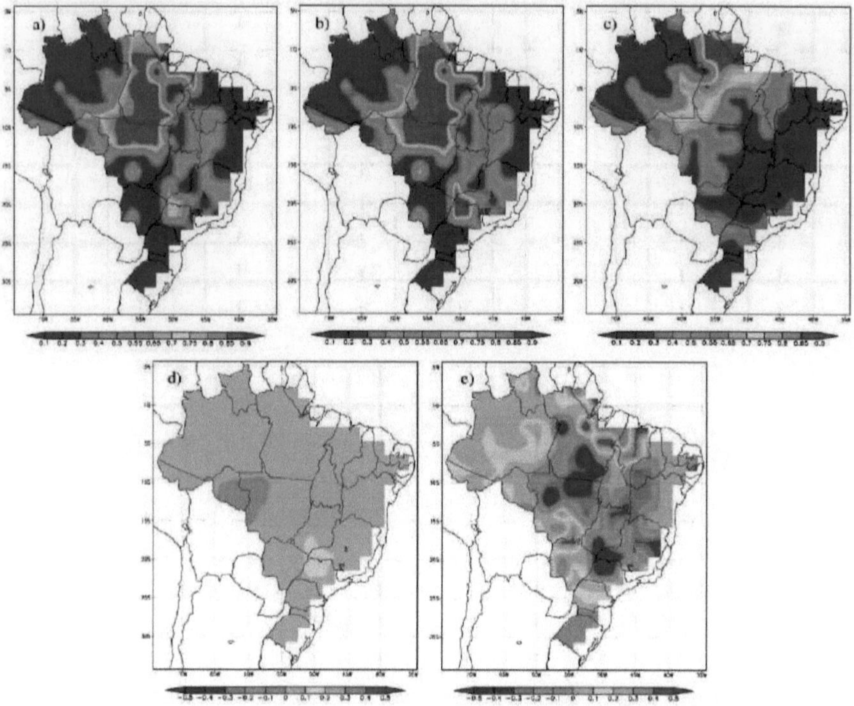

Figure 5.9: a) Setzer index (SI) for current climate and vegetation for September, b) SI for future climate and current vegetation for September, c) SI for future climate and vegetation simulations for September; d) are the anomalies between b) and a); e) are the anomalies between c) and a).

5.3.5 Analysis for October

With the start of the rainy season in most of Brazil, the most favorable atmospheric conditions for burning are located in the far north of the country (Fig. 5.10). It is in this region that the highest number of fires occur during this period (Fig. 4.19a, b). The analysis of IS follows, with reasonable precision, the anomalies of Tmax, URmin and PSE (Fig. 5.11). Critical IS values are only observed in the northern part of the state of Parâ. It is important to note that, although there are fires mainly in Maranhâo in October, as shown in Figures 4.19a,b the IS does not present a considerable risk.

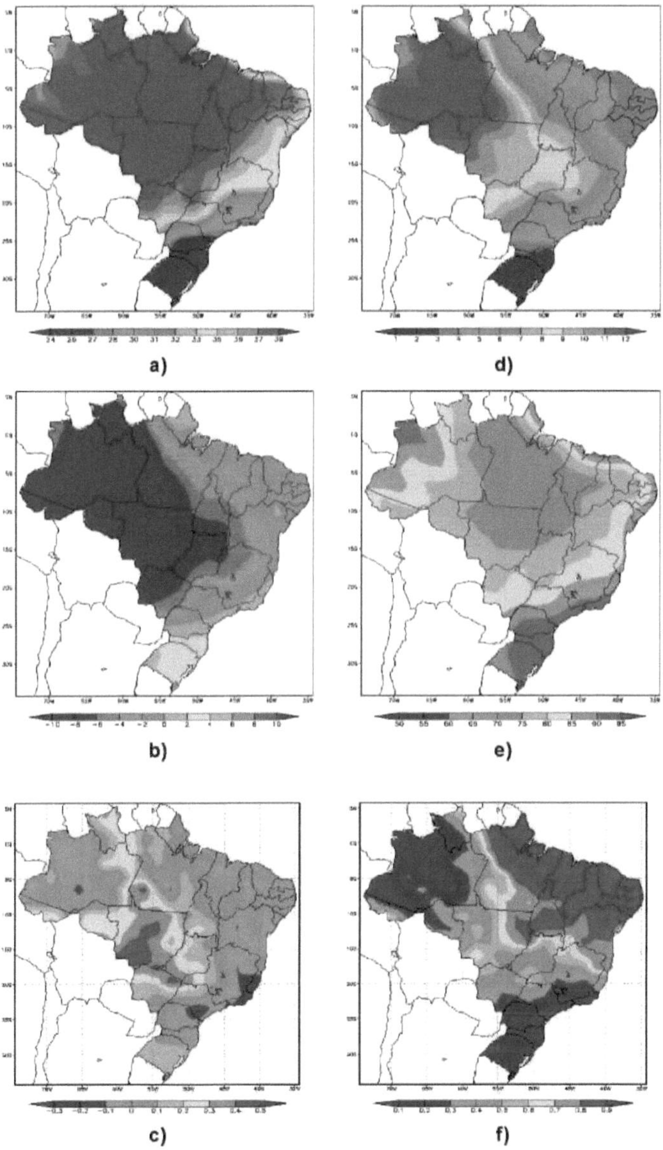

Figure 5.10: a) Tmax anomaly between the simulation of future climate and current vegetation and the present for October; b) the same as (a) for URmin and c) for PSE, d), e) ,f) shows the spatial distribution of Tmax, URmin and PSE with climate and vegetation simulations for October.

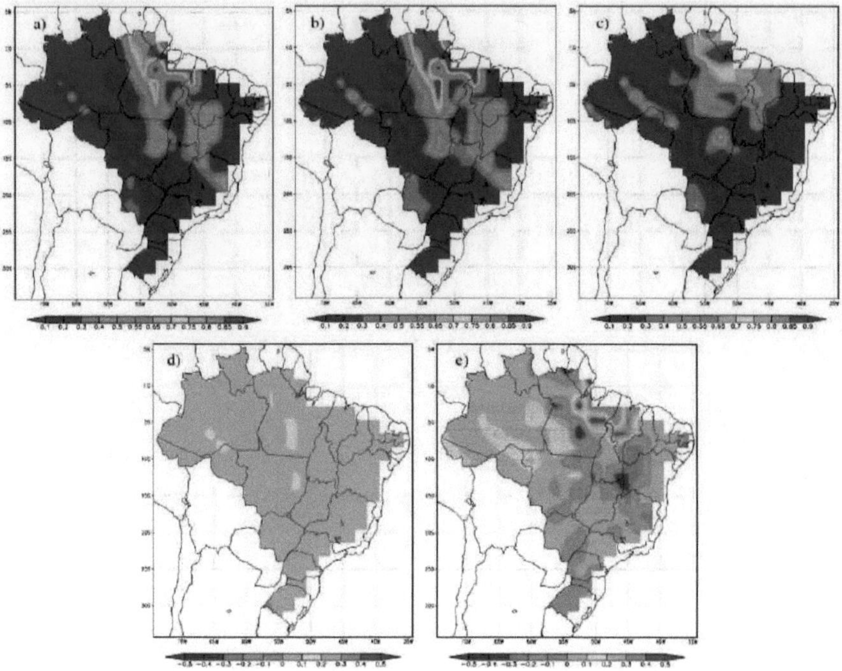

Figure 5.11: a) Setzer index (SI) for current climate and vegetation in October, b) SI for future climate and current vegetation in October, c) SI for future climate and vegetation simulations in October; d) are the anomalies between b) and a); e) are the anomalies between c) and a).

Chapter 6

6.0 Discussions and Conclusions

This study looked at possible climate impacts on the risk of wildfires in Brazil through long-term numerical simulations using the ECHAM5/MPI-OM Coupled Ocean-Atmosphere General Circulation Model, under current climate conditions and global warming. NCEP reanalysis data was used to validate the ECHAM5 results. The climate analyses for ECHAM5 and NCEP data showed that the model represents the current period relatively well in terms of temperature, precipitation and relative humidity, since it manages to capture the dominant climate patterns, particularly over the Amazon region. Over central Brazil, the model tends to overestimate temperature and rainfall, and underestimate relative humidity.

Based on the analysis of the risk of fires proposed by the Haines Index (HI) and the Setzer Index (SI), it was found that the main areas at risk of fires under current conditions are concentrated in the central-western and northern regions of Brazil for the months of June to October.

The comparison between IH and IS shows that IH has a lower fire risk index compared to IS. The indices also differ in terms of the regions most prone to large fires.

The calculation of the IS, based on two types of vegetation, one according to current conditions and the other as proposed for the future in a global warming scenario, shows that the future IS is extremely sensitive to the condition imposed by the vegetation, in the sense that the areas where the forest was replaced by cerrado had a greater risk of fires. On the other hand, areas where the savannah was replaced by grass had a reduced IS. This shows that, according to the anomalous atmospheric conditions predicted for the future, extremely favorable conditions can be expected for the development of large-scale fires, especially in the Amazon region. On the basis of the results presented, it can also be suggested that the replacement of the Amazon rainforest by savannah will lead to an increase in atmospheric susceptibility to fires.

presented, it can also be suggested that the replacement of the Amazon rainforest by savannah will lead to an increase in atmospheric susceptibility to fires.

7.0 BIBLIOGRAPHICAL REFERENCES

AGEE, J.K. Fire and pine ecosystems, in Ecology and biogeography of Pinus (ed. D.M. Richardson), **Cambridge University Press**, Cambridge, pp. 193-218, 1998.

ANDREAE, M.O. Biomass burning: Its history, use and distribution and its impact on environmental quality and global climate, in **Global Biomass Burning: Atmospheric, Climatic and Biospheric Implications**, edited by. Levine, J.S., pp. 3-21, 1991. MIT Press, Cambridge, Mass.

ANDREAE, M.O., ATLAS, E., CACHIER, H., COFER, W.R., HARRIS, G.W., HELAS, G., KOPPMANN, R., LACAUX, J.-P., WARD, D.E. Trace gas and aerosol emissions from savanna fires. **Biomass Burning and Global Change**, edited by J.S. Levine, pp. 278295, 1996a, MIT Press, Cambridge, Mass.

ANDREAE, M.O., ATLAS, E., HARRIS, G.W., HELAS, G., KOPPMANN, MAENHAUT., W., MANO, S., POLLOCK, W.H., RUDOLPH, J., SCHAEFFE, D., SCHEBESKE, G., WELLING, M. Methyl halide emissions from savanna fires in southern Africa. **Journal Geophys Res.,** v. 101, pp. 23603-23601, 1996b.

ANDREAE, M.o., RoSENFELD, D., ARTAXo, P., coSTA, A.A., FRANK, G.P., LoNGo, K.M., SILVA-DIAS, M.A.F. Smoking rain clouds over the Amazon. **Science**, v. 27, pp. 1337-1342, 2004.

ASLAN, Z., ToPÇU, S. Seasonal variation of surface fluxes and atmospheric interaction in Istanbul, September, pp. 22-27, 1994. Lisbon, Portugal.

AZZALI, S.; MENENTI, M. Mapping vegetation-soil-climate complexes in Southern Africa using temporal fourier analysis of NoAA-AVHRR NDVI Data. **International Journal of Remote Sensing**, v. 21, n 5, pp. 973-996, 2001.

BATISTA, A.C. Evaluation of controlled burning in *Pinus taeda L.* stands in northern Paranà. **Doctoral thesis,** Federal University of Paranà, Curitiba, 1995.

BoND, W.J., VAN WILGEN., B.W. Fire and plants population and community Biology series. **Kluwer Academic Publishers**, New York, 1996.

BRoTAK, T.L., REIFSNYDER, W.E. Predicting major wildfire occurrence. **Fire Management Notes**, v. 38, pp. 5-8, 1977.

BRoWN, I.F., SCHRoEDER, W., SETZER, A., MALDoNADo, M.J.R., PANToJA, N., DUARTE, A.F., MARENGo, J. Monitoring fires in Southwestern Amazônia rain forest. EoS,

Transaction American Geophysical Union, v. 87, n 26, pp. 253-264, 2006a.

BRoWN, I.F., MoULARD, E.M.N.P., NAKAMURA, J., SCHRoEDER, W., MALDoNADo, M.J.R., Vasconcelos, S.S., Serlhost, D. **Relatório preliminar do mapeamento de áreas de risco para incêndios no leste do Estado do Acre**. (UFAC/FUNTAC). Rio Branco, Brazil, p. 20, 2006b.

CHARNEY, J.G., QUIRK, W.J., CHOW, S.H., KORNFIELD, J. A comparative study of the effects of albedo change on drought in semi-arid regions. **Journal of Atmospheric Sciences**, v. 34, pp.1366-1385, 1977.

COCHRANE, M.A. Fire science for rainforests. **Nature,** v. 42, p. 913-919, 2003.

COOK, K., VIZY, E.K. Effects of Twenty-First-Century climate change on the Amazon rain forest. **Journal of Climate**, v. 21, n 3, p. 542-560, 2008.

COSTA, M.H., FOLEY, J.A. Combined effects of deforestation and doubled atmospheric CO2 concentrations on the climate of Amazonia. **Journal of Climate**, v. 13, pp. 18-34, 2000.

COSTA, M.H., YANAGI, SILVIA N.M, SOUZA, P.J.O.P., RIBEIRO, A., ROCHA, E.J.P. Climate change in Amazonia caused by soybean cropland expansion, as compared to caused by pastureland expansion. **Geophysical Research Letters**, v. 34, p. L07706, 2007.

CRUTZEN, P.J., ANDREAE, M.O. Biomass burning in the tropics: impact on atmospheric chemistry and biogeochemical cycles. **Science**, v. 250, pp. 1669-1678, 1990.

CRUZ, F.W., BURNS, S.J., KARMANN, I., SHARP, W.D., VUILLE, M., CARDOSO, A.O., FERRARI, J.A., DIAS, P.L.S., Jr, O.V. Insolation-driven changes in atmospheric circulation over the past 116,000 years in subtropical Brazil. **Nature**, v. 434, pp. 63-66, 2005. doi: 10.1038/nature03365

DIRMEYER, P.A., SHUKLA, J. The effect on regional and global climate of expansion of the world's deserts. **Quarterly Journal of the Royal Meteorological Society,** v. 122, pp. 451-482. 1996.

EMERY, W., THOMSON, R. Data analysis methods in physical oceanography. Pergamon Press, **Elsevier Science**, Oxford, p. 643, 1997.

FERRAZ-VICENTINI, K.R. History of fire in the cerrado: a palynological analysis. **Doctoral thesis**, University of Brasilia, (DF), 1999.

GRIMM, A. M. How do la Nina events disturb the summer monsoon system in Brazil? **Climate Dynamics**, v. 22, n. 2-3, pp. 123-138, 2004.

HAINES, D.A., WERTH, J. Haines Index climatology for the Westhern USA. **Westhern Region Technical Attachment**, v. 97, p. 17, 1987.

HAINES, D.A. A lower atmospheric severity index for wildland fires. **Natl. Wea. Dig.**, v. 13, pp. 23-27, 1988.

HAINES, T.K., MARTINEZ., J., CLEAVES., D.A. Influences on prescribed burning activity in the United States national forest system. **International Forest Fire News,** v. 19, pp. 43-46, 1998.

HINES, K.M., BROMWICH, D.H., MARSHALL, G.J. Artificial surface pressure trends in the NCEP-NCAR reanalysis over the Southern Ocean and Antarctica. **Journal Climate** v. 13, pp. 3940-3952, 2000.

KOCHTUBAJDA, B., FLANNIIGAN, M.D., GYAKYUM, J.R., STEWART, R.E. The influence of atmospheric instability on fire behavior in Northwest Territories. **Symposium on fire and forest Meteorology meeting**, 2001.

http://www.kni.ml/~sterl/Essence

INTERGOVERNMENTAL PANEL ON CLIMATE CHANGE (IPCC), Climate Change 2007. **The Physical Science Basis.**

JUNGCLAUS, J.H, KEENLYSIDE, N., BOTZET, M., HAAK, H., LUO, J.-J., LATIF, M., MAROTZKE, J., MIKOLAJEWICZ, U., ROECKNER, E. Ocean circulation and tropical variability in the coupled model ECHAM5/MPI-OM**, Journal Climate,** v. 19, pp. 39523972, 2006.

JUSTINO, F.B., SOUZA, S.S., SETZER, A. Relationship between hotspots and meteorological conditions in Brazil. **Proceedings of the XII Brazilian Congress of Meteorology.** 2002.

JUSTINO, F.B. The influence of boundary conditions on the Last Glacial Maximum**. Shaker Verlag**, v. 127, p. 16, 2004.

JUSTINO, F.B., TIMMERMANN, A., KREBS, U., SOUZA, E. Synoptic reorganization of atmospheric flow during the Last Glacial Maximum. **Journal Climate**, v. 18, pp. 28262846, 2004.

KALNAY, E., KANAMISTU, M., KISTER, R., COLLINS, W., DEAN, D., GANDIN, L., IREDELL, M., SAHA, S. WHITE, G., WOOLLEN, J., ZHU, Y., CHELLIAN, M., EBISUZAKI, W., HIGGINS, W., JANOWIAK, J., MO, K.C., Ropelewski. C., Wang. J., Leetman, A., Reynolds. R., Jenne, R., Joseph, D. The NCEP/NCAR 40-year reanalysis project. **Bulletin**

of the American Meteorological Society, v. 77, pp. 437471, 1996.

KAUFMAN, Y.J., TANRE, D., BOUCHER, O. A satellite view of aerosols in the climate system, **Nature**. 419.6903.215-223, 2002.

KIRCHOFF, V. SCAR-B proceedings, Transec Editorial, INPE, Sao José dos Campos, (SP), p. 208, 1997.

KLAR, A.E. Water in the soil-plant-atmosphere system. Sao Paulo, **Nobel**, p. 408, 1984.

KRIKYLA, K., HAMEED, S. Harmonic analysis of the seasonal cycle in precipitation over the United States: a comparison between observations and a general circulation model, **Journal of Climate**, v. 2, pp.1463-1475, 1989.

LATURNER, N., SCHERER, H.W. Burning and forest fires in Mato Grosso. **Revista Geonotas**, v. 8, 2004.

LEMOS, C.F. The Haines index as an indicator of the development of hotspots in Brazil through the Eta Regional Model. **PhD Thesis**, Fluminense Federal University, Rio de Janeiro, (RJ), 2006.

MARSLAND, S.J., HAAK, H., JUNGCLAUS, J.H., LATIF, M., ROSKE, F. The Max Planck Institute global ocean/sea-ice model with orthogonal curvilinear coordinates. **ocean. modell**, v. 5, pp. 91-127, 2003.

NEPSTAD, D.C., VERiSSIMO, A., ALENCAR, A., NOBRE, C., LIMA, E., LEFEBVRE, P., SCHLESINGER, P., POTTER, C., MOUTINHO, P., MENDOZA, E., COCHRANE, M., BROOKS, V. Large-scale impoverishment of Amazonian forests by logging and fire, **Nature**, v. 398, pp. 505-508, 1999.

NUNES, J.R.S. FMA^+ - A new forest fire danger index for the State of Paraná. **PhD Thesis**, Federal University of Paraná, Department of Forestry Engineering, 2005.

OLIVEIRA, P.H. O efeito dos aerossóis de queimadas sobre o fluxo de CO_2 na Amazônia. **Master's dissertation**, University of Sao Paulo, Sao Paulo, (SP), 2005.

OYAMA, M.D. Climatic consequences of vegetation change in northeastern Brazil: a modeling study. **PhD thesis,** National Institute for Space Research, Sao José dos Campos, (SP), 2002.

PAGE, S.E., SIEGERT, F., RIELEY, J.O., BOEHM, H.-D.V., JAYA, A., LIMIN, S. The amount of carbon released from peat and forest fires in Indonesia during 1997. **Nature**, v. 420, pp. 61-65, 2002.

PANOFSKY, H.A., BRIER, G.W. Some applications of statistics to meteorology. Pennsylvania. **University Park**, p. 224, 1968.

PHILANDER, G. El Nino Southern Oscillation (ENSO) Models. **Encyclopedia of Ocean Sciences**, pp. 827-832, 2001.

PINTO, L.I.C. Comparison of precipitation products and incident solar radiation for South America: observed data and reanalyses. **Master's Thesis**, Federal University of Viçosa, Viçosa, (MG), 2007.

PROCÓPIO, S., ARTAXO, P., KAUFMAN, Y.J. REMER, L.A. SCHAFER, J.S., HOLBEN, B.N. Multiyear analysis of Amazonian biomass burning smoke radiative forcing of climate. **Geophysical Research letters**, v. 31, n 3, 2004. Article No. L03108.

RAMANATHAN, V., CRUTZEN, P.J., KEIL, J.T., ROSENFELD, D. Atmosphere: aerosols, climate, and the hydrological cycle. **Science**, v. 294(5549), pp. 2119-2124, 2001.

RIBEIRO, G.A., BONFIM, M,V.R. Incêndio florestal versus queimima controlada. **Revista Açao Ambiental**, ano II, n 12, pp. 8-11,2000.

RIBEIRO, G.A.; SOARES, R.V. Characterization of forest combustible material and effects of controlled burning on its reduction in a Eucalyptus viminalis stand. **Cerne,** Lavras, v. 4, n 1, pp. 57-72, 1998.

ROECKNER, E., BAUML, G., BONAVENTURA, L., BROKOPF, R., ESCH, M., GIORGETTA, M., HAGEMANN, S., KIRCHNER, I., KORNBLUEH, L., MANZINI, E., RHODIN, A., SCHLESE, U. SCHULZWEIDA, U., TOMPKINS, A. The atmospheric general circulation model ECHAM5. Part I: Model description. Max Planck Institute for Meteorology Rep. 349, 127 pp. [available from MPI for Meteorology, Bundesstr. v. 53, p. 20146, 2003. Hamburg, Germany].

SALDARRIAGA, J.G., WEST, D.C. **Quaternary Research**, pp. 26-53, 1986.

SALGADO-LABOURIAU, M.L., FERRAZ-VICENTINI, K.R. Fire in the Cerrado 32,000 years ago. **Current Research in the Pleistocene**, v. 11, pp. 85-87, 1994.

SANFORD, R.L., SALDARRIAGA, J.G., CLARK, K.E., UHL, C., HERRERA, R. **Science**, pp. 277-358, 1985.

SETZER, A., PEREIRA, M.C., PEREIRA Jr, A.C. The use of NOAA satellites in the detection of fires in Brazil. **Climanàlise**, v. 7, n 8, pp. 40-53, 2002.

SETZER, A., SISMANOGLU, R.A. Fires in Brazil. **Climanàlise Boletim de Monitoramento e Anàlise Climàtica**, CPTEC/INPE, v. 21, n. 01, p. 1,2006.

SISMANOGLU, R.A., SETZER, A., JUSTINO, F., LIMA, W.F.A. Initial evaluation of the performance of the fire risk generated by CPTEC. **Proceedings of the XII Brazilian Congress of Meteorology,** 2002.

SUD, Y.C., FENNESSY, M.J. A study of the influence of surface albedo on July circulation in semi-arid regions using the GLAS GCM. **Journal of Climatology,** v. 2, pp.105-125, 1982.

SUD, Y.C., FENNESSY, M.J. Influence of evaporation in semi-arid regions on the july circulation: a numerical study. **Journal of Climatology,** v. 4, pp.383-398, 1984.

VERA, C., SILVESTRI, G., LIEBMANN, B., GONZALEZ, P. Climate change scenarios for seasonal precipitation in South America from IPCC-AR4 models. **Geophysical Research Letters**, v. 33, n 13, 2006. doi:10.1029/2006GL025759.

WHITMORE, T.C. An introduction to tropical rain forests**. Oxford University Press**. Oxford, 1990.

WILKS, D.S. Statistical methods in the atmospheric sciences: an introduction. San Diego, California, **Academic Press**, p. 467, 1995.

WINKLER, J.A., POTTER, B., WILHELM, D., SHADBOLT, R., BIAN, X., PIROMSOPA, K. A cimatology of the Haines Index for North America derived from NCEP/NCAR reanalysis fields. Sixth Fire and Forest Meteorology Symposium, **American Meteorological Society** (2005).

XUE, Y., SHUKLA, J. The influence of land surface properties on Sahel climate. Part I: Desertification. **Journal of Climate**, v. 6, pp. 2232-2245, 1993.

XUE, Y., SHUKLA, J. The influence of land surface properties on Sahel climate. Part II: Aforestation. **Journal of Climate**, v. 9, pp. 3260-3275, 1996.

YAMASOE, M., VON RANDOW, C., MANZI, A., SCHAFER, J., ECK, T., HOLBEN, B. Effect of smoke on the tramissivity of photosynthetically active radiation inside the canopy. **Atmospheric Chemistry Physics Discuss**, 5550-5934, 2005,

MIX
Papier aus verantwortungsvollen Quellen
Paper from responsible sources
FSC® C105338

Printed by Books on Demand GmbH, Norderstedt / Germany